Mete[illegible]
Meteorites

origins and observations

Meteors and Meteorites

origins and observations

Martin Beech

The Crowood Press

First published in 2006 by
The Crowood Press Ltd
Ramsbury, Marlborough
Wiltshire SN8 2HR

www.crowood.com

British Library Cataloguing-in-Publication Data
A catalogue record for this book is available from the British Library.

ISBN 1 86126 825 4
EAN 978 1 86126 825 9

Frontispiece: This bright Perseid meteor was captured during a 15-minute guided camera exposure. The bright star to the left of the meteor trail is Epsilon Eridani. (Image courtesy the European Southern Observatory)

Dedicated to my wife, Georgette

Typeset by Focus Publishing, 11a St Botolph's Road, Sevenoaks, Kent

Printed and bound by Biddles Ltd, King's Lynn

Contents

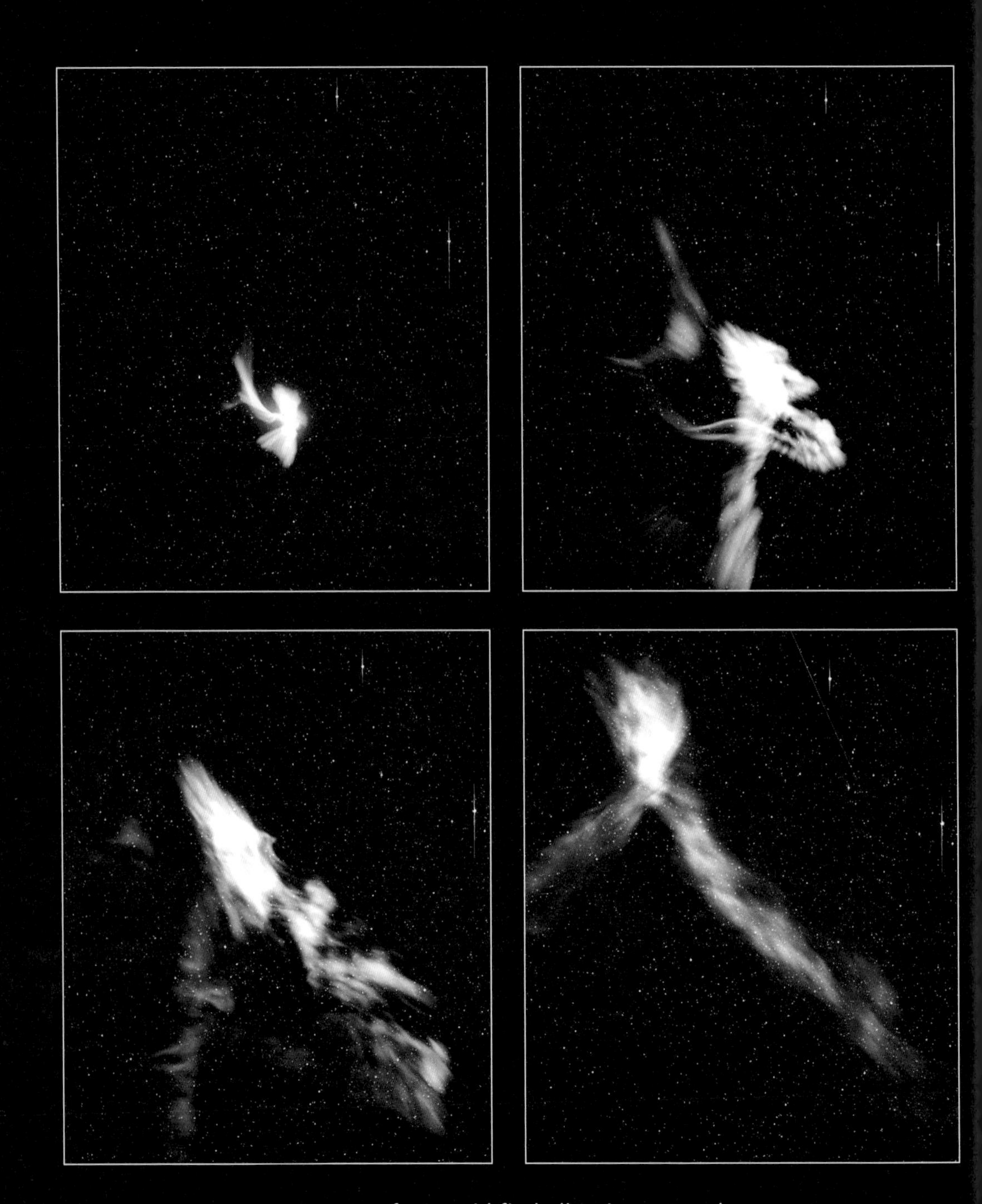

Images of a Leonid fireball train captured on the night of 17 November 1998.
(See page 54 for further details)

Preface

Meteors have been observed throughout recorded history, and they have been known by many different names. Shooting stars, falling angels, stars in wane, licking flames, and flying torches – all attempts to name and capture the essence of the streaks of transient light that scythe through the darkness of the night sky with startling speed and grace. They are objects of wonderment indeed. Historically associated with approaching death by some cultures and with the release of souls from purgatory by others, the humble shooting star could bring the observer good luck, or it could be an omen of ill-fortune. Very bright meteors, or fireballs, were known as draco volans, literally 'flying dragons'. These fiery drakes would occasionally drop their stolen treasures – as stones with centres of solid gold – to the ground. In some cultures it was believed that the strange black rocks and masses of iron that occasionally fell from the sky, amid great atmospheric concussions and rumblings, must have special powers.

Meteors and meteorites, steeped in folklore, objects of ancient myths and legends, distorted facts and abundant fiction, have inspired the human imagination. And in the modern era, as this book will hopefully reveal, meteors and meteorites are still objects of great wonderment when viewed in the context of our knowledge of them as phenomena of the solar system. We now know that shooting stars are derived from ageing comets (themselves viewed in the past with great fear and misunderstanding) and that meteorites are the fragments ejected into space when asteroids collide in the deep expanse between Mars and Jupiter. The modern-day story of the origins of meteors and meteorites is no less amazing than the folk-tales of the ancient past, and their observation and study is no less enthralling.

My hope in writing this book is that you will be inspired to go out and observe meteors. And perhaps by reading and learning more about the physics and astronomy of meteors, meteorites and meteoroid streams, you will even be inspired to make scientifically useful observations of meteors and to search for meteorites. The early chapters of this book are a little heavy on science and terminology, but I urge you to persevere with the details: they are important, and even if the poets among us might beg to differ, there is great beauty in the present-day understanding of meteors and meteorites. Enjoy the physics, enjoy the phenomena, enjoy the hunt, and enjoy the observing – but most of all, make your observations count.

Martin Beech
November 2005

1 The Particulate Sea

The solar system contains many millions of objects. By far the biggest and most impressive of these objects is the Sun. With a diameter 109 times that of the Earth and a mass 333,000 times greater, it is the Sun by its gravitational attraction that literally holds the entire solar system together. Next in order of size and dominance are the eight major planets (Mercury to Neptune), which move around the Sun on orbits with radii between 0.4 and 30AU (astronomical units). From the orbit of Neptune extending outwards along the plane of the ecliptic to distances of several hundred astronomical units are the Kuiper Belt objects. This ancient collection of ice-worlds, of which Pluto, Sedna and the recently discovered object temporarily designated 2003 UB313 (which is possibly bigger than Pluto), are the largest currently known, delineate the outer remnants of the solar nebula – the disk out of which the Sun and the solar system formed some 4.56 billion years ago. Still farther out, at distances of many tens of thousands of astronomical units from the Sun, we encounter the Oort Cloud, an ancient reservoir of cometary nuclei that enshrouds the solar system in a vast, near-spherical envelope of slowly moving icy bodies. The Oort Cloud marks the very boundary of all that constitutes the solar system. Beyond the Oort Cloud, the Sun's gravitational influence no longer holds sway, and it is the rest of the Galaxy and the nearby stars that determine the outcome of orbital motion.

For all its vast size and volume, the solar system is far from being a quiet place. Born out of the tumultuous collisions of massive planetesimals – the first large objects to form in the solar system's disk, the rocky planets and satellites, even to this very day, betray the scars of countless batterings and cosmic catastrophes. Likewise, the main-belt asteroids, confined to the region between Mars and Jupiter, and the Kuiper Belt objects are all collisionally evolved and evolving: there is a perpetual grinding-down of these objects through violent encounters and by the continual strafing of minor impacts. And, by a series of intricate gravitational pirouettes, the fragments produced by collisions in the Kuiper Belt are channelled past the planets in the outer solar system to be fed into the family of short-period comets whose orbits are controlled by the gravitational pull of Jupiter. Brought then into orbits that carry them close to the Sun, these cometary nuclei become heated and baked, giving off sweeping tails composed of small silicate grains. These grains are our future shooting stars, or meteors. Likewise, the stately motions of comets in the Oort Cloud are occasionally thrown into turmoil by close encounters with stars or with the giant gas clouds that pervade the disk of our Milky Way galaxy. Careening past the Sun with a rapid burst of motion, these long-period comets spill their ancient cargo of small stony grains into the inner solar system, adding still further to the reservoir of potential meteors. Indeed, it is from the midst of this perpetual chaos of encounters, collisions and impacts that a 'particulate sea' of small silicate grains to

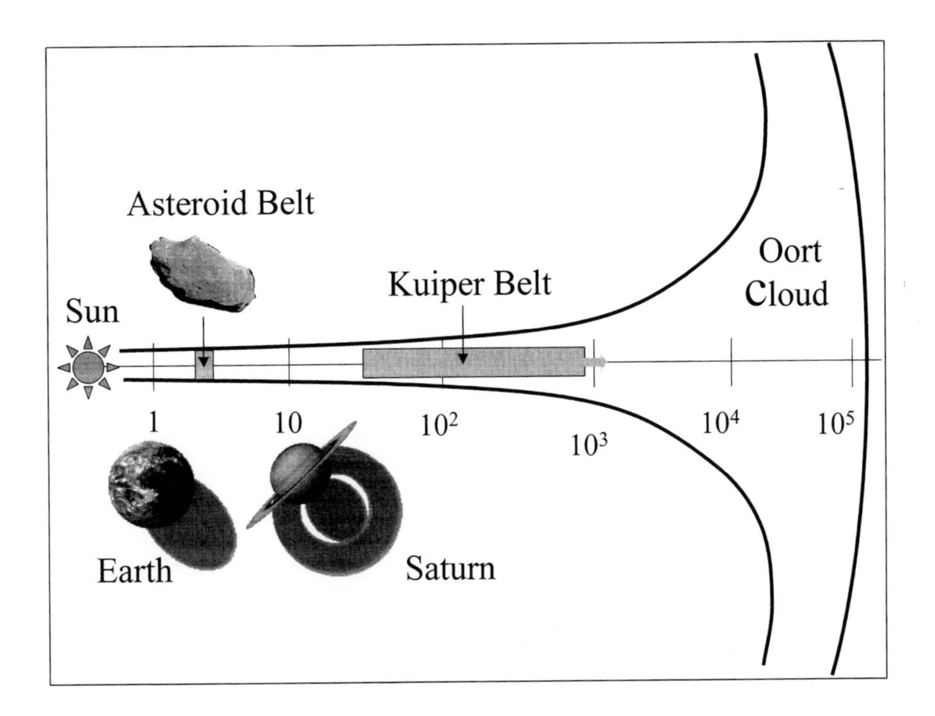

Fig. 1 *A schematic vertical cross-sectional view of the solar system. The numbers below the central line (which is in the plane of the ecliptic) correspond to distances from the Sun in astronomical units, and the scale is a logarithmic one. The Earth is 1AU from the Sun, while Saturn is at 9.54AU. The planets, asteroids and Kuiper Belt objects move around the Sun in orbits whose inclinations are close to the plane of the ecliptic. Beyond several thousand astronomical units from the Sun, however, the orbital distribution becomes more spherical (shown as a flaring-out of the boundary lines). The outer boundary of the solar system, just beyond* 10^5*AU from the Sun, is shown by the curved arc to the far right of the diagram. The nearest star to the Sun (Proxima Centauri) is about three times farther away than the outer Oort Cloud boundary, at a distance of some* 3×10^5*AU.*

metre-sized fragments has arisen – a sea of particles that literally washes over and envelops the entire inner solar system.

We can experience directly the existence of the particulate sea, and the Earth's perpetual journey through it, for on any clear night of the year we can see the intermittent and fleeting passage of shooting stars. We have all seen these wayward meteors as they dash headlong across the celestial vault, leaving in their wake a faint scratch of lingering light. Indeed, the appearance of a shooting star indicates a journey's end for at least one component of the particulate sea: the encounter with the Earth's atmosphere typically destroys the particle long before it can reach the ground. Meteorites, fragmentary meteoroids produced through the ponderous collisions of asteroids, are the exception to this rule. And yet, while they are not fully destroyed by their passage through

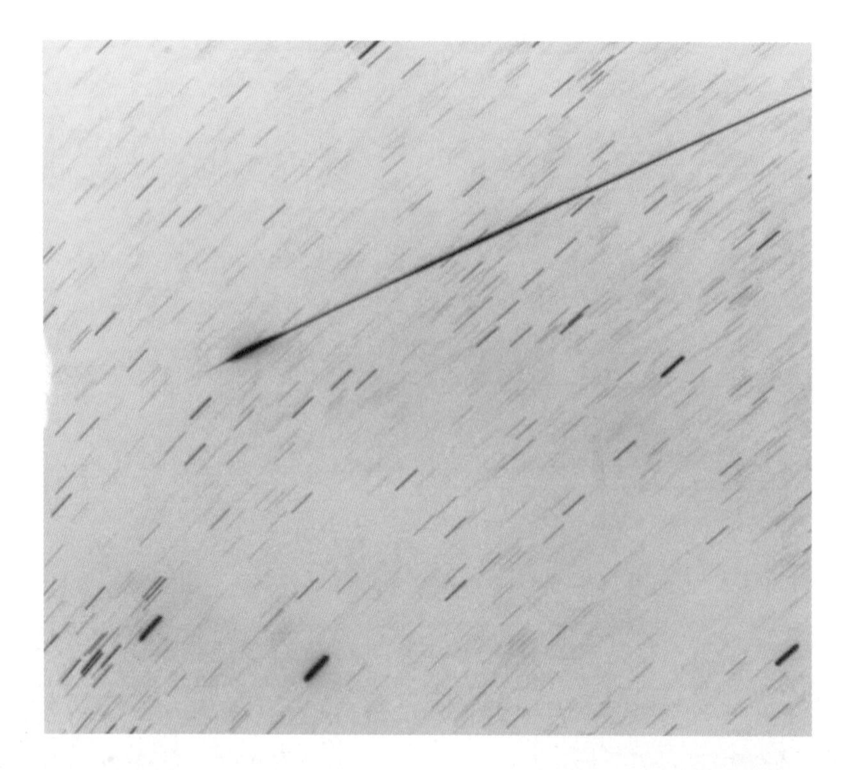

Fig. 2 (above) *A Leonid meteor streaks through the constellation Taurus. The meteor lasted for about half a second and was captured during a 5-minute star-trail exposure. This image, like others in this book, is shown as a negative to reveal more detail. (Original image by John Williams, http://leonids.hq.nasa.gov/leonids/)*

the Earth's atmosphere, the arrival of these cosmic rocks on the surface of our planet does represent an end in the sense that their days of 'surfing' the particulate sea are over.

A Few Formal Definitions

The physicist Steven Toulmin once quipped, 'Definitions are like belts. The shorter they are the more elastic they need to be.' In this vein, definitions concerning meteors tend to be short and consequently they carry a number of caveats. The International Astronomical Union (IAU) set out a series

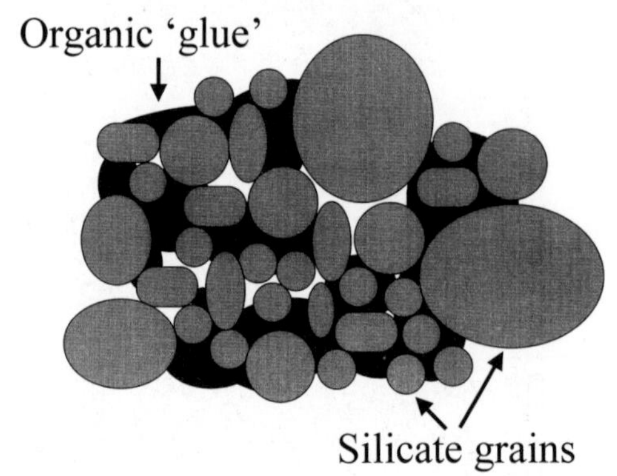

Fig. 3 *No actual meteoroid capable of generating a visual meteor has ever been collected or photographed outside the Earth's atmosphere. The schematic image here depicts a so-called dust-ball model meteoroid in which the meteoroid is envisaged as being composed of numerous small grains, of varying size and shape, all held together by an 'organic glue' mixture of low melting point.*

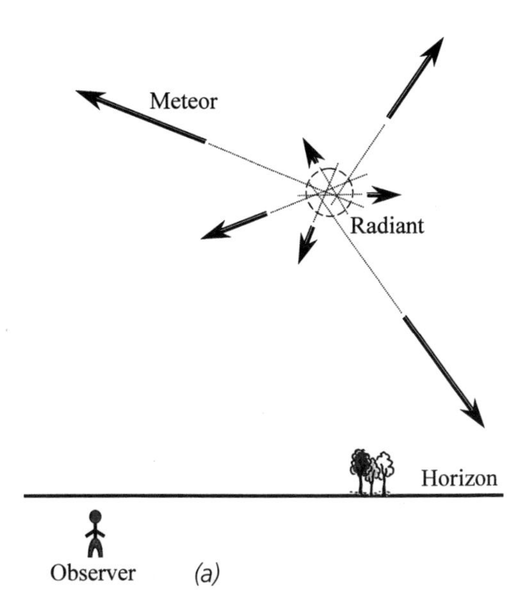

of nomenclature guidelines for meteor astronomy in 1961. The subcommittee charged with the standardization of terms carefully distinguished between physical and observable characteristics, and in this respect a **meteor** is defined as 'the light phenomenon which results from the entry into the Earth's atmosphere of a

Fig. 4 *(a) Meteors associated with a particular meteor shower all appear to radiate from a small region on the sky – the radiant. The radiant can be located by tracing the observed meteor paths (solid arrows) backwards (thin lines) to a common intercept. (b) A series of superimposed video frames reveal the radiant location of the Quadrantid meteor shower. (Image courtesy Sirko Molau)*

Fig. 5 Allan Hills 84001 – a stony meteorite found in Antarctica. This meteorite is actually material blasted from the surface of the planet Mars, and is a sample of that planet's crust. It has been suggested that the meteorite contains 'fossilized' microbial organisms – but the evidence for this has been hotly debated. (Image courtesy NASA)

solid particle from space'. A **meteoroid**, on the other hand, is 'a solid object moving in interplanetary space, of a size considerably smaller than an asteroid and considerably larger than an atom or molecule'. So, while a meteor is observable, a meteoroid, generally, is not. Building upon these definitions, we may immediately correct the much perpetuated malapropism that the solar system contains 'meteor streams' – it does not, but it most certainly does contain **meteoroid streams**. A **meteor shower** is seen when the Earth passes through a meteoroid stream, and such showers may take place annually or intermittently according to the age and characteristics of the meteoroid stream encountered. A **meteorite** is defined as a meteoroid that 'has reached the surface of the Earth without being completely vaporized'.

The IAU nomenclature subcommittee also developed working definitions for the terms **micrometeorite** and **dust**, the former being defined as a 'meteoritic particle with a diameter in general less than a millimetre', and the latter as 'finely divided solid matter, with particle sizes in general smaller than micrometeorites'. The appearance of the meteor that results when a meteoroid enters the Earth's upper atmosphere depends strongly upon the initial size of the meteoroid. For this reason a range of observing techniques are required to record the various meteors produced by different sizes of meteoroids. Dust particles, for example, produce no observable phenomena – visual or otherwise – whereas a meteorite-producing event will be presaged by a brilliant fireball easily visible to the naked eye. (A **fireball** is defined as 'a bright meteor with luminosity which equals or exceeds that of the brightest planets' – which in effect means brighter than magnitude –4.7, the greatest brilliancy that Venus can attain.) For a meteoroid to produce a meteor – a phenomenon detectable by optical means or by radar – upon entering the Earth's atmosphere, it must be larger than about 100 micrometres in size (1 micrometre = 10^{-6}m, written as 1μm). A meteoroid smaller than about 100μm can lose its kinetic energy – its energy of motion – in the form of thermal radiation rather than mass loss, and consequently such objects produce no distinct light or ionization trail. Although dust particles produce no meteoric light, they do scatter sunlight, and the dust distributed around the plane of the ecliptic can be seen with the naked eye just before sunrise or after sunset, under dark-sky observing conditions, as the zodiacal light.

***Fig. 6** A high-magnification image of a micrometeorite spherule, about 10μm across. (Image courtesy NASA)*

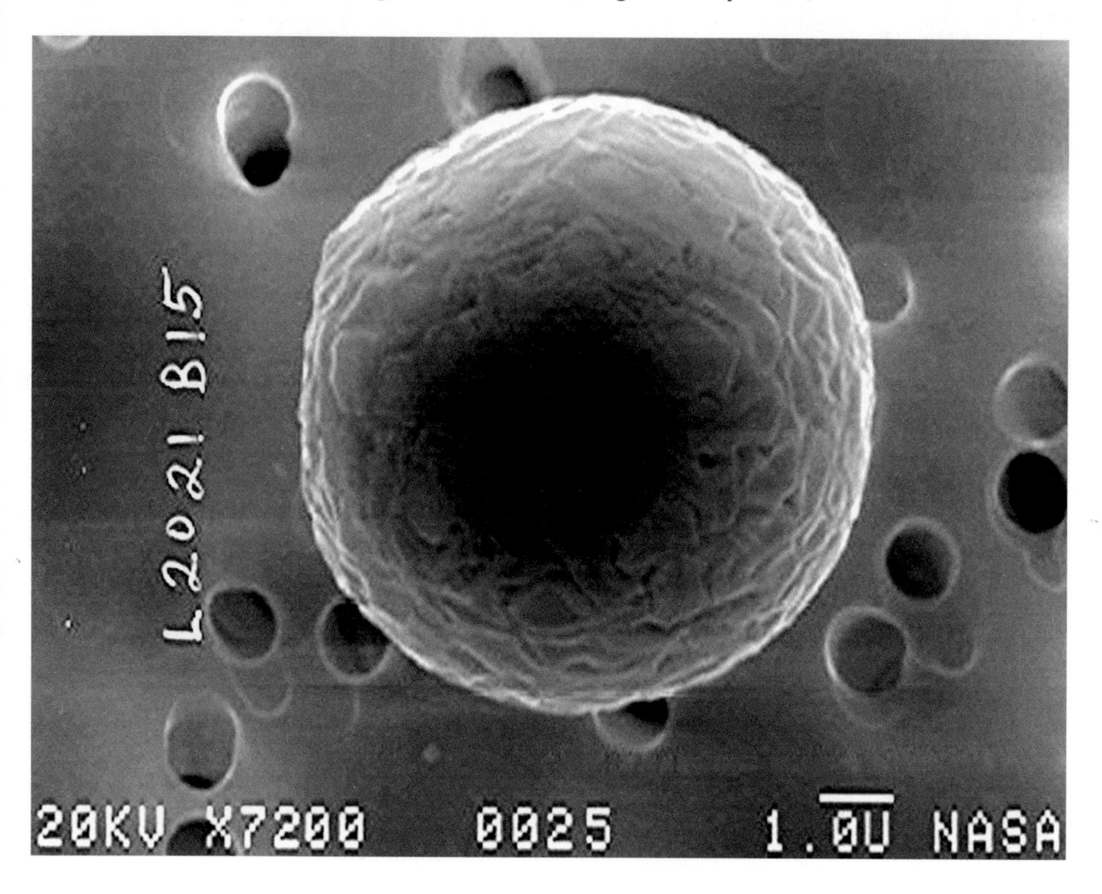

Fig. 7 (below) *An interplanetary dust particle (IDP). Note the open, 'fluffy' structure. The cosmic spherule shown in Figure 6 probably looked something like this IDP (only larger) before it melted upon entering the Earth's atmosphere. This particle was collected during a high-altitude aircraft flight and is about 100µm in size. (Image courtesy NASA)*

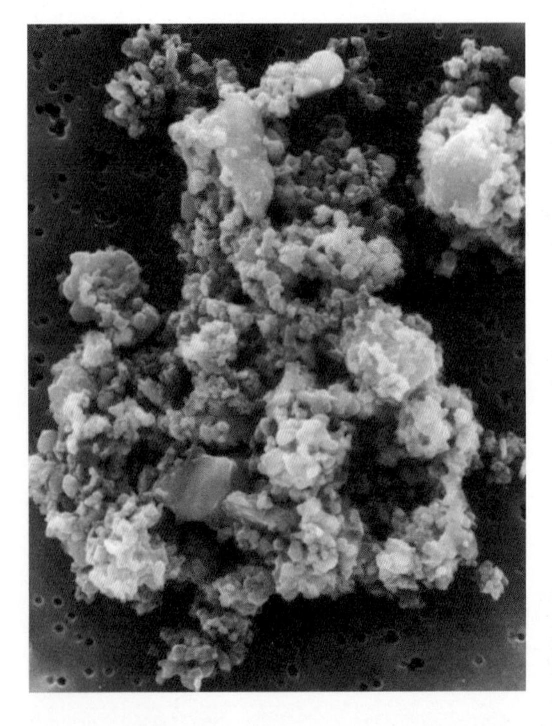

At the opposite end of the size scale to dust particles, there is no formally recognized limit above which a large meteoroid becomes a small asteroid. Sir William Herschel introduced the word **asteroid** into the astronomy lexicon early in the nineteenth century to describe the star-like appearance of the newly discovered minor planets that circle the Sun between Mars and Jupiter. The key observational characteristic of an asteroid is that it is large enough to reflect enough sunlight to produce a star-like image in a telescope. A **comet**, on the other hand, is recognized observationally by its possession of a tail and coma. All this being said, whether an asteroid is recognized for what it is depends ultimately upon the efficiency of the telescope used to observe it, and that of course changes with time. In the modern era, asteroids as small as 5 to 10m in diameter are routinely detected in the region

inside the Moon's orbit, and that would appear to be a reasonable boundary line beyond which large meteoroids become small asteroids.

Pulling together all the definitions given above, we may now define the size range of the objects we are concerned with in this book:

Fig. 8 *An exceptionally bright fireball recorded on 21 January 1999 with an all-sky camera system located in the Czech Republic. (Image courtesy Dr Pavel Spurny, Astronomical Institute, Ondrejov Observatory)*

Fig. 9 The asteroid Ida and its moon, Dactyl, imaged by the Galileo spacecraft in 1993. The asteroid is 56km long; Dactyl is about 1.5km across. Asteroids are predominantly composed of silicates and nickel–iron metal. (Image courtesy NASA)

we are interested in solid objects moving through space, with a size larger than 100mm but less than 5m, that can potentially intercept the Earth's orbit.

Unravelling the Meteoric Phenomenon

Meteors are an upper-atmosphere phenomenon: the light emitted when a meteor sweeps across the sky is generated typically at altitudes between 115 and 85km. The largest

Fig. 10 Comet West as seen on 9 March 1976. This particular comet displayed a broad dust tail and a small but bright coma. Cometary nuclei are predominantly made of water ice. (Image courtesy NASA and John Laborde)

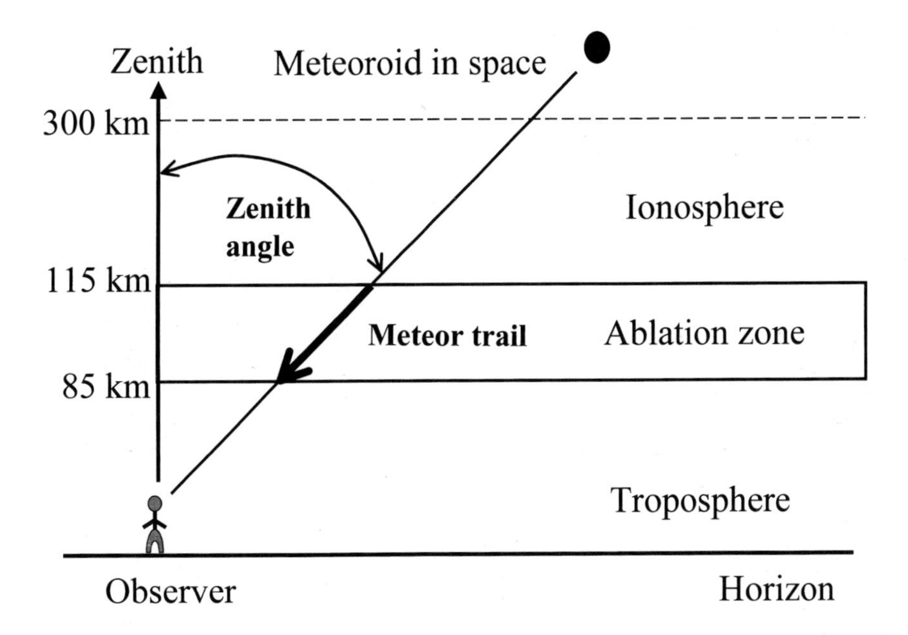

Fig. 11 *The meteor region of the Earth's atmosphere. Typical naked-eye meteors ablate at altitudes of between about 115 and 85km. Meteoroid heating begins at about 300km. The ionosphere (used in short-wave radio communications) begins at about 140km. The troposphere, where our everyday weather is generated, extends to about 15km above ground level. Also indicated in the diagram is the zenith angle, which is the angular distance from the observer's zenith to the start of the meteor trail.*

meteoroids, those that produce bright fireballs, can penetrate more deeply into the Earth's atmosphere, depending upon their initial speed. If large enough, a meteoroid can survive its entire passage through the Earth's atmosphere to land as a meteorite.

Measurements of initial meteoroid speeds show that they vary widely, from a minimum of 11.2km/s to a maximum of 72.8km/s. The lower limit is related to the Earth's escape velocity, and is the velocity an object would acquire if it simply fell from deep space towards the Earth under the Earth's gravitational attraction alone. The upper limit corresponds to a head-on collision between the Earth at perihelion, where its orbital velocity is 30.3km/s, and a meteoroid having the solar system's escape velocity of 42.5km/s at 0.98AU (the Earth's perihelion distance) from the Sun.

The maximum brightness of a meteor is directly related to the initial mass and velocity of its associated meteoroid. Specifically, it is the meteoroid's kinetic energy ($KE = \frac{1}{2}mv^2$, where m is the initial mass and v is the initial speed) that determines whether it is completely destroyed in the Earth's upper atmosphere or will survive, to reach the ground, as a meteorite. In general, massive and fast meteoroids produce bright meteors, while small and slow meteoroids produce faint meteors.

When a meteoroid begins to encounter molecules in the Earth's upper atmosphere, at about 300km up, it becomes heated through

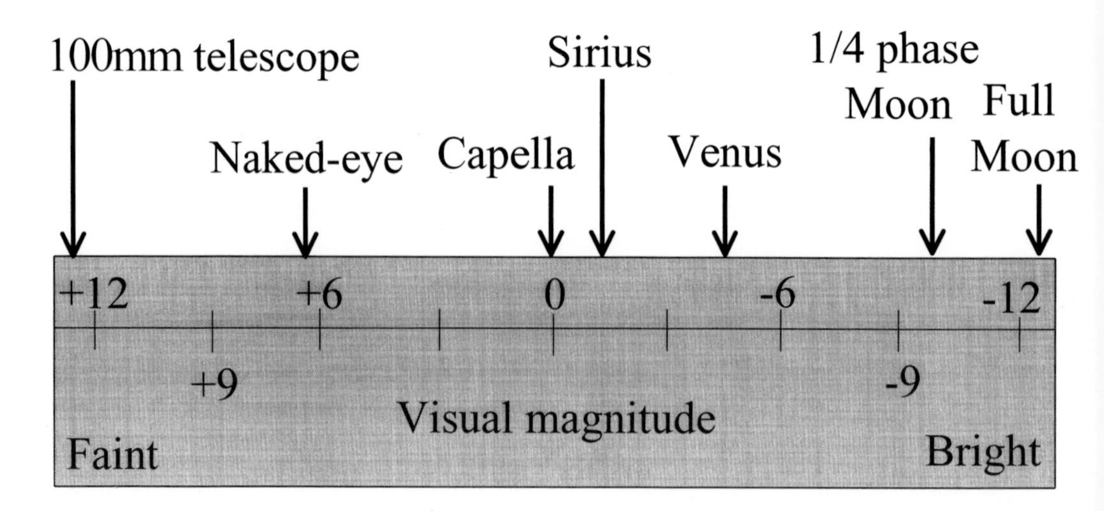

Fig. 12 *The magnitude scale for various solar system and stellar objects.*

direct surface collisions – essentially, the kinetic energy of any colliding molecules is converted into heat energy at the surface of the meteoroid. The meteoroid will continue to heat up until a surface temperature of about 2,000°C is reached. Once this temperature has been achieved, typically at an altitude of about 115km, material will begin to ablate from the meteoroid's surface, producing a trailing wake of electrons and atoms in an excited or ionized state. At this stage the meteoroid is beginning to lose mass and is also beginning to gradually slow down. Collisions between meteoric ions and atmospheric atoms result in the production of a plasma trail behind the meteoroid. The excited atoms in the plasma trail will eventually lose energy through the emission of photons, and the electrons will also eventually recombine with the ions to produce more photons. It is the emission of these photons that produces a meteor's light. Spectroscopic observations of meteor trails indicate that it is mainly the atoms and ions from the meteoroid that produce the observed light.

Meteor brightness is typically expressed as a magnitude. The apparent visual magnitude m_V is defined as $m_V = -2.5 \log \text{flux} + C$, where flux corresponds to the energy received in the visual part of the electromagnetic spectrum per second per unit area, and C is a constant. The absolute visual magnitude M_V of a meteor is defined as the apparent visual magnitude corrected to a standard height of 100km at the observer's zenith. The manner in which the magnitude scale is defined results in bright objects having negative magnitudes and faint objects having large, positive magnitudes. The Sun, for example, has an apparent visual magnitude of –26.75, and the full Moon has an apparent visual magnitude of –12.7, while Venus at maximum brightness reaches magnitude –4.7. The faintest stars visible to the human eye, under ideal observing conditions, are about magnitude +6.5. The faintest star that can be seen through a telescope (its limiting magnitude) is conveniently calculated by the formula

$$m_V = 2.7 + 5 \log D$$

where D is the diameter of the telescope's aperture measured in millimetres. A telescope with a 100mm diameter objective will therefore typically reveal stars to a limiting magnitude of about +12.5.

In addition to depending upon the initial meteoroid mass and velocity, a meteor's brightness will also depend upon its **zenith angle**, Z (as shown in Figure 11). This is simply the angular distance, measured in degrees, from the start of the meteor trail to the observer's local zenith, the point immediately overhead. If $Z = 0°$ then the meteor trail will be moving directly downward. The most likely observed zenith angle will be $Z = 45°$. A commonly used relationship between the maximum brightness m_V, the meteoroid's initial speed v, its initial mass m and the zenith angle Z was derived by Luigi Jacchia (Smithsonian Astrophysical Observatory) and co-workers, who in the early 1950s conducted an extensive photographic study of meteor trails. The formula/equation (1) is

$$m_V = 4.84 - 2.25 \log m - 8.75 \log V - 1.5 \log(\cos Z)$$

The units in equation (1) are kilograms for the meteoroid mass and kilometres per second for the velocity. The zenith angle is measured in degrees, and the magnitude is the apparent magnitude at maximum brightness. Equation (1) reveals that for a zenith angle of 45°, a zero-magnitude meteor (i.e. one as bright as the star Capella – *see* Figure 12) will be produced by a 6g (or 20mm diameter) meteoroid when the initial speed is 15km/s. Increase this speed to 35km/s, and a zero-magnitude meteor will be produced by a 0.2g (10mm diameter) meteoroid. And for an initial speed of 65km/s, to produce a zero magnitude meteor the mass of the meteoroid need only be 0.02g (3mm diameter). The variation of maximum apparent

Fig. 13 *Meteor magnitude plotted against initial speed, based upon equation (1). The curves are labelled according to the initial mass of the meteoroid in kilograms, and in each case the zenith angle is taken to be 45°.*

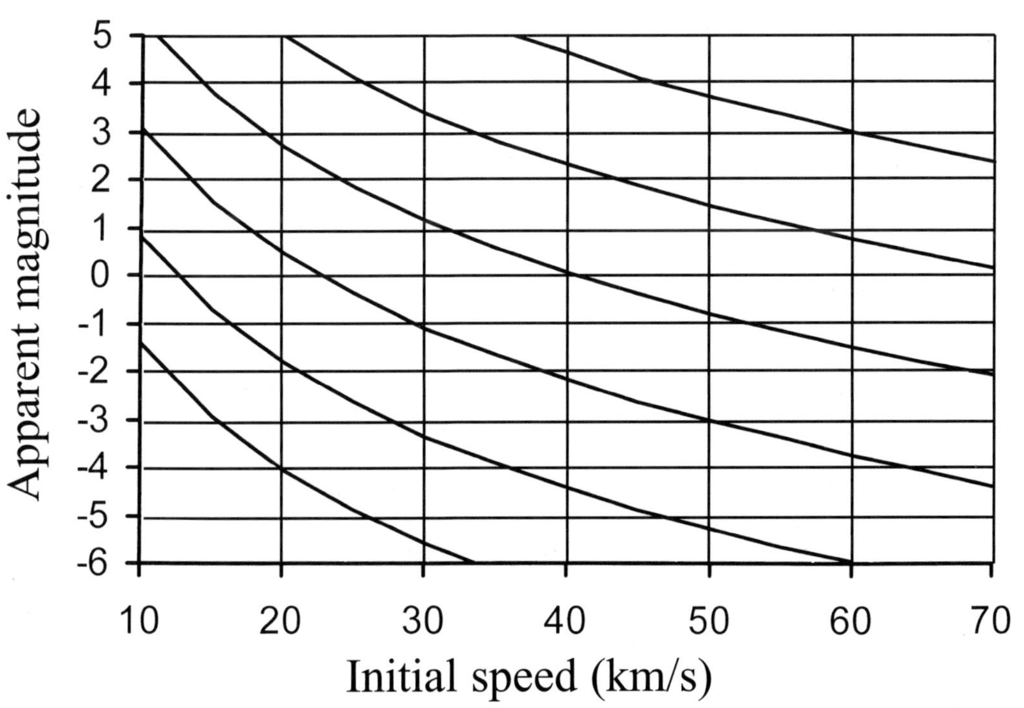

magnitude with velocity for a selection of initial meteoroid masses is shown in Figure 13.

If it is assumed that a meteoroid is a solid, spherical mass, then its diameter d and mass m are related via the density δ of meteoroidal material. By definition, the density is the mass divided by the volume, so for a spherical meteoroid the diameter is

$$d = 2\left(\frac{3m}{4\pi\,\delta}\right)^{1/3} \qquad (2)$$

Unfortunately meteoroid densities are only poorly determined, and consequently there is some uncertainty in the δ term in equation (2). That said, the available data indicate that typically δ ~ 800kg/m^3, but there is also a high-density subset of meteoroids with δ ~ 3,000 to 8,000kg/m^3. Figure 14 shows a representative diagram of meteoroid size versus mass. Clearly, from equation (2), small size accompanies small mass; and, as implied by equation (1), the smaller the mass of a meteoroid (for a given speed), the fainter the maximum brightness of the meteor produced. This variation dictates a series of 'natural' limits to the observational techniques that can be applied to the study of meteors. The human eye, for example, can 'see' meteors only if they are brighter than magnitude +6.5, and consequently visual observers are restricted to recording those meteors produced by meteoroids larger than about 1mm in diameter or more massive than 10^{-6}kg. Fireballs, by definition, are meteors brighter than approximately magnitude –5, and hence all-sky camera systems designed to monitor fireball activity will register only those meteors produced by objects larger than about 50mm across or with masses greater than

Fig. 14 *Meteoroid size in metres plotted against meteoroid mass in kilograms. The scales are logarithmic, and the upper magnitude scale corresponds to an initial speed of 25km/s and a zenith angle of 45°. It has also been assumed here that the density of meteoroidal material is 1,000kg/m^3. The dashed line corresponds to equation (2).*

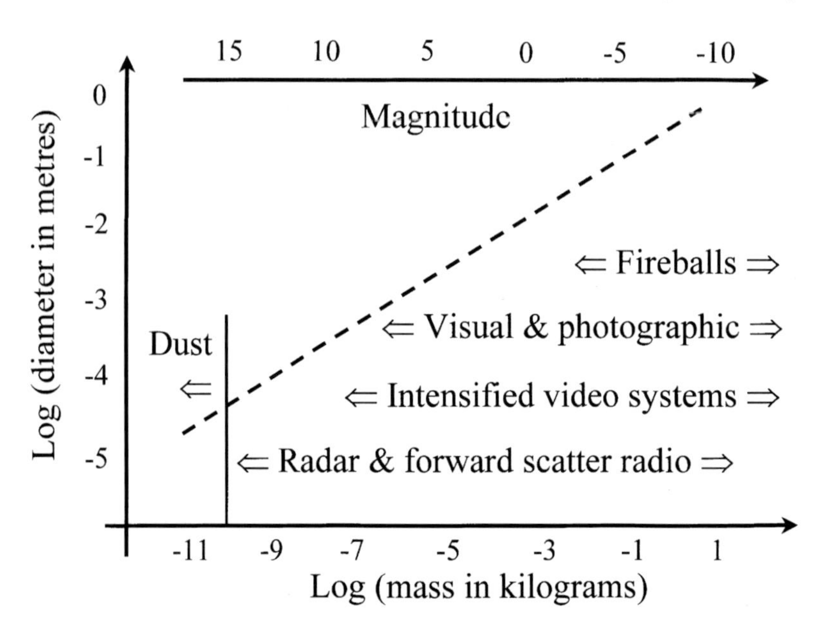

0.1kg. Enhanced video and low-light-sensitivity television cameras can record meteors as faint as magnitude +9, and are detecting meteors produced by meteoroids larger than about 0.5mm across or more massive than about 10^{-7}kg. And finally, radar and forward-scatter systems (*see* Chapter 7) can detect meteors as small as 0.1mm across, with masses of about 10^{-10}kg. Meteoroids smaller than about 100µm will not produce an ionization trail and consequently do not qualify as meteor-producing bodies.

Meteor Power

In the previous section I introduced the expression for kinetic energy – the energy a body has by virtue of its motion. While energy is expressed in joules, power is expressed in joules per second, or watts. The power or energy dissipated by a meteor can be expressed in terms of the initial kinetic energy of the meteoroid, $\frac{1}{2}mv^2$, divided by its duration, ΔT. If, for example, a 1mm diameter meteoroid (i.e. one with a mass of about 5×10^{-7}kg)

Fig. 15 *The Barringer impact crater. This 1.2km diameter crater was formed about 50,000 years ago when an asteroid some 50 to 100m across crashed into the Arizona desert. (Image courtesy the US Geological Survey)*

encounters the Earth with an initial speed of 35km/s and the resultant meteor lasts for 0.25 seconds, then the energy dissipated in terms of the meteor's power will be

$0.5 \times 5 \times 10^{-7} \times (35 \times 1{,}000)^2/0.25 = 1{,}225$ watts

This result tells us that the energy released in the destruction of a 1mm diameter meteoroid is about the same as is required to 'power' a 100W light bulb for 12.25 seconds. Alternatively, a 10mm diameter meteoroid with an initial speed of 35km/s, producing a meteor lasting for, say, 0.5 seconds, will have a power of about 0.6 million watts. The energy dissipated by this particular meteor, therefore,

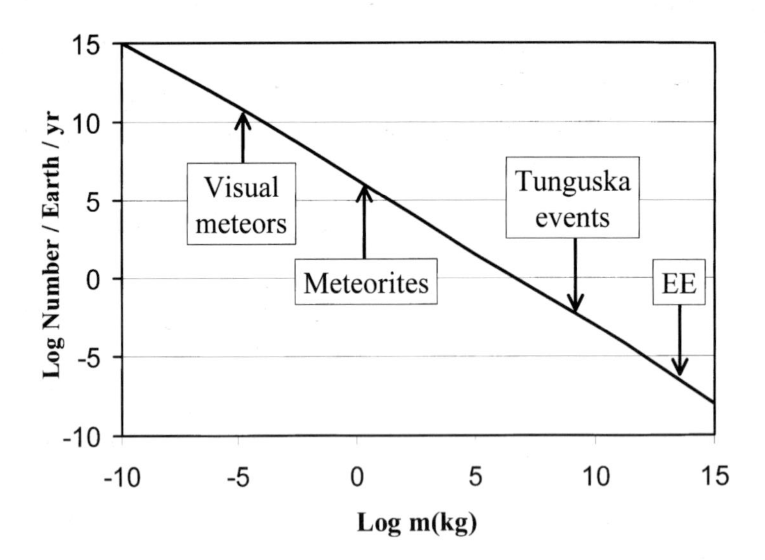

Fig. 16 The total number of objects N of mass larger than a specific mass falling to Earth per year. While extinction events (EE) are the most devastating in their effect on the environment, it is evident from the diagram that they are very rare, one such impact occurring on average every few tens of millions of years.

is the same as is required to run a 100W light bulb for about 107 minutes. Equation (1) indicates that such a 10mm diameter meteoroid will produce a meteor of about magnitude –1 peak brightness.

Leonid fireballs carry prodigious amounts of kinetic energy as a consequence of their very high initial speeds, of about 70km/s. One of the brightest fireballs to be observed during the Leonid storm in the year 2000 reached a peak brightness of magnitude –14.4, endured for some 0.74 seconds, and was produced by a meteoroid with an estimated mass of 1kg. With these characteristics, the initial kinetic energy of the fireball's underlying meteoroid was some 2.45 billion joules, which is equivalent to the explosive energy of 585kg of TNT. The energy dissipated by the fireball amounted to 3.3 billion watts, which is enough to run a 100W light bulb for 383 days.

Fig. 17 The flattened Tunguska forest in Siberia. This photograph, obtained by Dr Leonid Kulik, was taken 19 years after an asteroid with an estimated diameter of 50m broke apart before it hit the ground on 30 June 1908. The resultant airburst explosion produced no impact crater but caused widespread devastation. (Image courtesy NASA)

The Accumulation of Cosmic Matter

As the Earth ploughs through the particulate sea it continually accretes matter. By observing meteors, gathering samples of meteorites, collecting cosmic dust from geological strata of different ages, and studying large impact craters, an estimate of the number of meteoroids of various masses encountering the Earth each year can be made. The total number of objects larger than a given mass falling to the Earth's surface is illustrated in Figure 16.

Luckily for our everyday existence, large asteroid impacts upon the Earth's surface occur

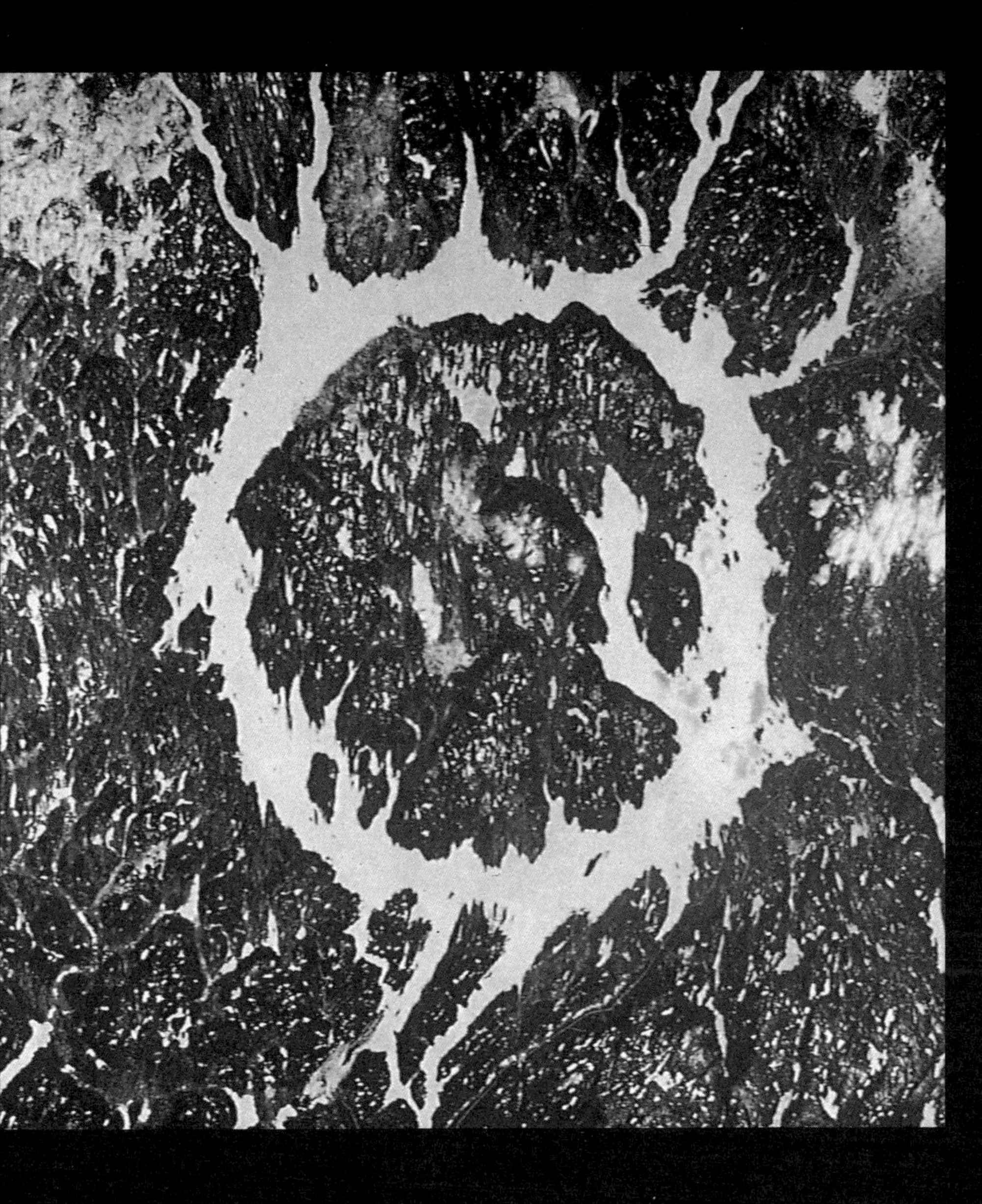

***Fig. 18** The Manicouagan impact crater in Quebec, Canada. This crater, now mostly collapsed and partially filled with water, is about 100km across and was formed by a large asteroid impact 212 million years ago. The circular lake actually defines the crater's central uplift. (Image courtesy NASA)*

only rarely. This is evident from Figure 16, which shows, for example, that the approximate impact frequency for objects several tens of metres across (with masses of about 10^8kg) is about one every five hundred years or so. The last known 'impact' with an object many tens of metres across was at Tunguska, Russia, on 30 June 1908. While no crater was produced in this airburst event, some 2,000km^2 of Siberian forest was destroyed. Energy equivalent to about five megatons of TNT was liberated in the Tunguska explosion, and the blast wave it produced circled the entire Earth many times. Very large impacts are capable of significantly altering the Earth's climate and destroying extensive ecosystems. The last such major impact occurred some 65 million years ago and ushered in the demise of the dinosaurs

Measurements made from spacecraft indicate that in the size range of about 20 to 500μm, some 30 million kilograms of extraterrestrial material is intercepted by the Earth each and every year. The total (long-term) average influx of material over the mass range from about 10^{-15}kg to 10^{15}kg incident upon the Earth's surface amounts to something like 150 million kilograms a year, much of this mass being accounted for by very large impacts which occur only rarely. Meteoroids that produce meteors are close to the first peak in the mass influx diagram (*see* Figure 19), at 10^{-8}kg, which partly explains why meteors are so commonly seen. Meteorite-producing meteoroids with masses between a few grams and a few thousand kilograms, on the other hand, constitute a relatively small fraction of the total influx. This latter result is important since it is through the study of meteorites that we learn about the chemistry, origin and physical structure of both extraterrestrial material and the early solar system – we indeed learn a lot from a very little.

The Accumulation of Meteorites

Determining just how many meteorites fall to Earth each year is not a trivial task. Indeed, one of the biggest problems associated with the collection of meteorites is that water covers most of the Earth's surface and consequently the majority of meteorites that fall are lost to us. However, since about 1970 various camera surveys have operated across Europe, Canada and the United States and have enabled astronomers to estimate how many meteorites

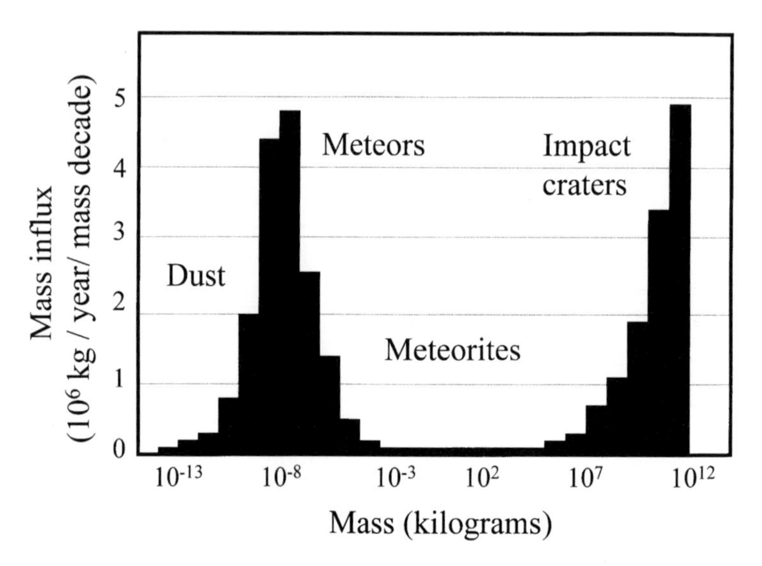

Fig. 19 *The long-term average mass influx for material encountering the Earth. The x-axis is divided into 'mass decades' – the mass range in each division increases by a factor of ten for each step to the right – and the mass influx is expressed in units of millions of kilograms per year per mass decade.*

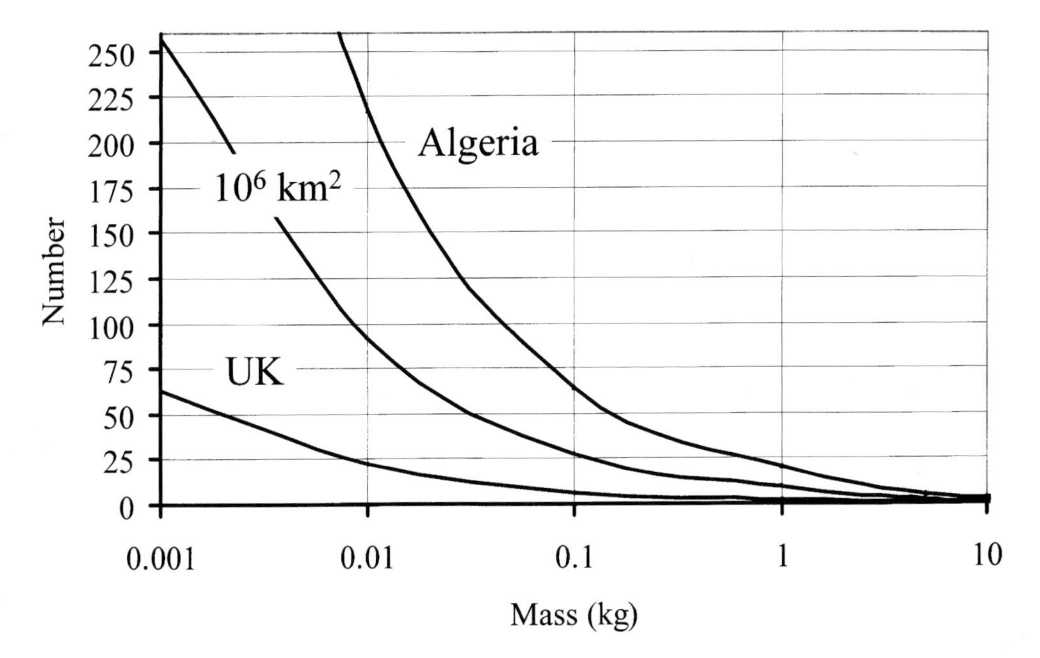

Fig. 20 *The cumulative number of meteorite falls with masses greater than M (kg) per year for the UK, Algeria and a land mass of area one million square kilometres. The cumulative number of falls in Algeria is given since it is from this hot desert region that many new meteorite finds are now being made. (Based upon the MORP study by Halliday, Blackwell and Griffin,* Meteoritics Journal, *vol. 24, p. 173, 1989)*

---

come to ground each year. One of the best studies on meteorite fall statistics was based on data from the Meteorite Observation and Recovery Project (MORP) conducted in Canada from the early 1970s to the mid-1980s. The results from MORP are in the form of the cumulative number (designated N) of meteorites more massive than a given mass that fall in a specified area each year.

Specifically, the total number of meteorites with masses greater than M, where M is greater than 1kg, falling per year per million square kilometres is given by

$$\log N = -0.82 \log M + 0.94$$

For meteorites less massive than 1kg the relationship is slightly different:

$$\log N = -0.49 \log M + 0.94$$

Using these formulae, we can determine that something like 257 meteorites larger than 1g fall each year over an area of one million square kilometres. Now, the surface area of the United Kingdom is about a quarter of a million square kilometres, so on average something like 62 meteorites larger than 1g fall in the UK each year. The MORP survey results further indicate that something like two meteorites larger than 5kg fall per million square kilometres each year. For the UK that translates into a meteorite more massive than about 5kg falling once every 2 to 3 years. Figure 20 shows the expected number of falls larger than a given mass for the UK and Algeria, and for a land mass of one million square kilometres.

The Dangerous Skies

If meteorites are continually falling from the sky, what is the probability that one might hit a car or a person? Such impacts must be rare because we don't read about them in newspapers very often, but they do happen. Meteorites have struck at least four cars in the past fifty years, and several people have received glancing blows from space rocks – but no one, as far as can reasonably be verified, has ever been killed by a meteorite. Many buildings have been damaged by meteorite impacts throughout history, and there are several instances of ships being struck. So, what are the actual chances of someone being killed by a meteorite?

To calculate the probability of being struck by a meteorite, we need to know the total area of the human target and the flux of meteorites (i.e. the number of meteorites falling per square kilometre per year). The total number of human beings on Earth in 2004 was estimated to be 6.4 billion. If we assume that each human being occupies an area of 0.25m^2, then the total human target area worldwide amounts to some 1,600km^2. Now, let us assume that each human being spends on average 2 hours a day outside (in the open, outside a building or car). The MORP survey results indicate that some 257 meteorites more massive than 1g fall to Earth each year per million square kilometres. So in an area of 1,600km^2 we would expect 0.4 meteorites more massive than 1g to fall in a year. The average 2 hours per day exposure time for each person amounts to a total exposure time of about 730 hours per year (or equivalently, $\frac{1}{12}$ of a year per year of exposure time). The probability of someone in the world being hit by a meteorite larger than 1g is therefore 0.4 × ($\frac{1}{12}$) = 0.03 per year. In other words, one person (out of the entire world's population) should statistically be hit by a meteorite once every 1/0.03 = 33 years. The likelihood of an individual who lives to, say, the ripe old age of 99 years being hit by a meteorite is therefore about 3 divided by 4.6 billion, or one chance in 1.5 billion. At these odds it is most definitely not worth taking out any insurance against being struck by a meteorite. Mrs Annie Hodges of Sylacauga, Alabama, USA is the only person in well-documented history to have been significantly injured by a falling meteorite. This happened on 28 November 1954, when she was struck a blow in the abdomen, while resting on a couch, by a 4kg meteorite which had punched a hole through the roof of her house. This story reminds us that even very low-probability events can, and indeed do, occur.

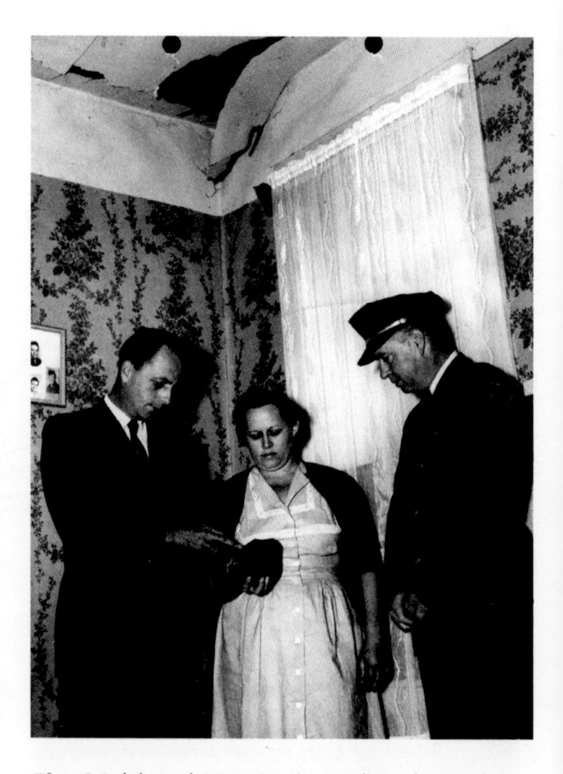

Fig. 21 *(above) Mrs Annie Hodges (centre), along with the mayor of Sylacauga (left) and the city's police chief (right), standing below where the meteorite crashed through the ceiling of her house. (Image courtesy the University of Alabama Museum of Natural History)*

Fig. 22 A far-infrared image of the sky based upon data gathered by the Cosmic Microwave Background Explorer (COBE) satellite. The image shows the entire infrared sky, with the centre of the Milky Way galaxy located at the centre of the image. The dust that delineates the disk of our galaxy is seen as the bright band across the middle of the image. The sideways, S-shaped swathe cutting right through the centre of the image is caused by zodiacal dust particles that reside in the ecliptic plane of the solar system. (Image courtesy the COBE Project and NASA)

Interstellar Meteoroids

Not all the meteoroids that produce detectable meteors in the Earth's atmosphere originate within the solar system. A small fraction, perhaps 1 per cent, of the total number of very faint meteors recorded at the Earth's orbit are produced by 'alien' meteoroids. These interlopers have entered the solar system from the depths of interstellar space, and amazingly, some may even be small fragments of planets, asteroids or comets that formed around other stars.

Astronomers have known for many decades that small dust grains permeate the space between the stars. Indeed, it was first realized nearly a century ago that starlight is systematically reddened with increasing distance from the Sun by the interstellar dust located in the disk of our Milky Way galaxy. The reddening of starlight comes about because the interstellar grains are just the right size to preferentially scatter blue light. This effect sets the grain size to be about 10^{-7}m – the same size as the wavelength of blue light. The interstellar dust grains are mostly pieces of 'stardust' emitted in the stellar winds of both very young and very old stars. Just as soot forms on a cool surface placed above a candle flame, as a stellar wind expands and cools, so solid particles can condense out of the gas phase, and it is these small graphite and silicon grains that seed the interstellar medium.

The key to recognizing an interstellar meteoroid is the measure of its initial speed upon entering the Earth's atmosphere. If the observed initial speed is greater than 72.8km/s, then it must have entered the solar system from deep space. Probably no interstellar meteor has ever been observed by the unaided human eye – the grains are simply not large enough to produce meteors as bright as magnitude +6.5. However, interstellar meteors are routinely detected by detectors on spacecraft and by Earth-based radar and image intensified video systems.

Radar monitoring of interstellar meteor activity over the past decade has shown that there is a strong seasonal variation in the arrival rate. In particular, Jack Baggaley of the University of Canterbury, New Zealand, and his colleagues have identified two strong annual sources of interstellar meteoroids. One source peaks in early February, and the other peaks in mid-June (*see* Figure 23). The February source is apparently related to a flow of neutral helium gas through the solar system – the gas and dust being derived from young, hot stars in the direction of the Scorpius–Centaurus star-forming region. The mid-June source is believed to result from the 'headwind' produced by the solar system as it moves through our galaxy. Other sources that have been suggested for interstellar meteoroids are the young star Beta Pictoris, which has a surrounding ring of gas and dust, and the Geminga supernova event, which occurred about 650,000 years ago at a distance of some 230 light years from the Sun.

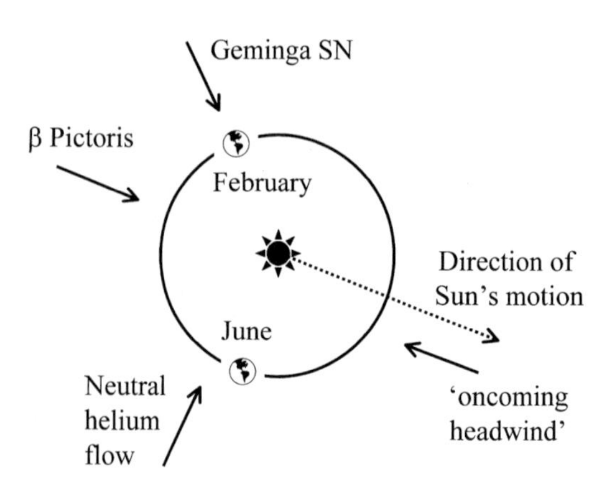

Fig. 23 *The approximate locations of interstellar meteoroid streams. The circle shows the Earth's orbit, and the two positions of the Earth correspond to the months in which the major peaks in the count rate of interstellar meteors are recorded. (After Taylor, Baggaley and Steel,* Nature, *vol. 380, p. 323, 1996)*

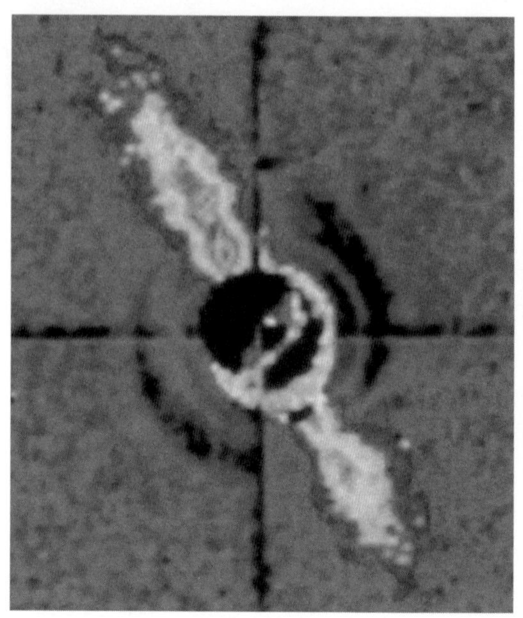

Fig. 24 *The star Beta Pictoris and its associated disk of dust (seen here edge-on). The star is located about 50 light years from the Sun, and may be actively forming planets within the disk at the present time. The disk is composed of small dust and ice grains and extends some 300AU either side of the parent star. (Image courtesy ESO)*

2 The View from Here

In its annual course around the Sun, the Earth continually encounters meteoroids. Day and night, the accretion never stops. There is literally always a meteor in the sky somewhere. But it is not just within the Earth's atmosphere: the Voyager 1 spacecraft, by good fortune, recorded the passage of a bright fireball through Jupiter's upper atmosphere on 5 March 1979. And a meteor was captured in the Martian skies by a camera aboard the Spirit robotic rover on 7 March 2004. Indeed, meteoroids strike all the planets and their attendant satellites in the solar system all the time. Dust detectors carried by spacecraft record hits from very small dust grains throughout their trajectories, no matter in which direction they go and no matter how far away from the Sun they are. In this chapter, however, our interests will be purely parochial, and we shall concentrate on the view as seen from the Earth.

Fig. 25 *While the Spirit Mars rover photographed the first Martian meteor,* Opportunity *(Spirit's* *twin companion) discovered the first meteorite on Mars. This image shows the basketball-sized nickel–iron meteorite found by* Opportunity *on 6 January 2005. (Image courtesy NASA/JPL/Cornell)*

Annual Activity

It might at first seem surprising that, in all of the chaos and randomness that produces the particulate sea, there should be any structure and order in the numbers of meteoroids encountered by the Earth – but order there most definitely is. Figure 26 shows the average number of meteors detected on each day of the year, over a five-year period, with a radar system stationed just outside Ottawa, Canada. It can be seen that the data do not fall at random in the diagram. The average daily number of meteors detected, for example, is lower in the first half of the year and higher in the second half. The data also reveal that short-duration activity spikes repeat themselves at intervals corresponding to exactly a year. The variation seen in Figure 26 is explained in terms of a sporadic background and the occurrence of annual meteor showers. The sporadic background accounts for the overall level of activity observed throughout the year (low in the first half of the year, higher in the second),

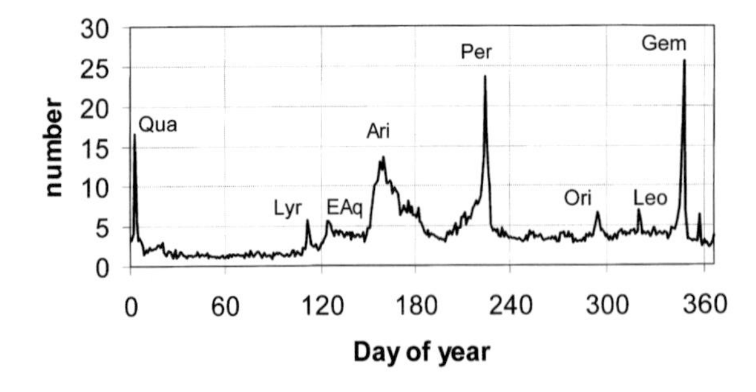

Fig. 26 *Daily averages of radar meteor echoes as recorded in Ottawa, Canada over the period from 1958 to 1962. The data are for meteors that were brighter than approximately zero visual magnitude. A number of meteor showers (*see *Table 1) are identifiable in the figure: the Quadrantids (Qua), the Lyrids (Lyr), the Eta Aquarids (EAq), the Arietids (Ari), the Perseids (Per), the Orionids (Ori), the Leonids (Leo) and the Geminids (Gem). (Data from Millman and McIntosh,* Canadian Journal of Physics, *vol. 42, p. 1730, 1964)*

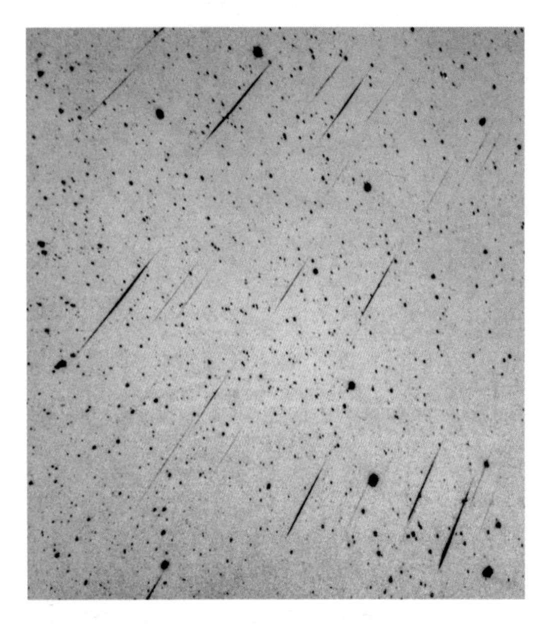

Fig. 27 *Leonid meteors moving along nearly parallel paths. (Picture courtesy NOAO)*

while the meteor showers account for the short-duration spikes seen at very specific times of the year.

Meteoroids from the sporadic background enter the Earth's atmosphere from all possible angles and with a range of speeds, whereas the meteoroids from a specific stream enter the Earth's atmosphere from the same direction in space and with the same initial speed (*see* Figure 27). The direction of entry and the speed of the various meteoroids will vary, however, from one meteor shower to another.

It's About Time

Before I consider the sporadic and meteor shower components in more detail, I shall first make a few comments about the measurement of time. Rather than identifying, for example, the time of year at which a shower reaches its maximum as the month, day and hour, meteor astronomers generally use the quantity known as solar longitude. The reasons for this are actually sensible and straightforward, but they require a little getting used to. **Solar longitude**, represented by the symbol $\lambda_\odot$ and expressed in degrees, is a precise measure of the location of the Earth in its orbit about the Sun, the time interval of one year corresponding to 360° of solar longitude. The zero point from which solar longitude is measured on the celestial sphere is the vernal (or spring) equinox (*see* Figure 28).

The reason that meteor astronomers use solar longitude rather than a calendar date to describe time is that there is an annual six-hour shift in the civil calendar year with respect to the actual solar year, culminating in the 29 February leap-year 'jump' every four years. There is no such yearly drift in the solar longitude, which defines the actual location of the Earth in its orbit. For example, the Perseid meteor shower reaches its maximum activity, in the sense of the greatest number of meteors visible per hour, at a solar longitude of $\lambda_\odot$ = 140°. The corresponding time, in the civil

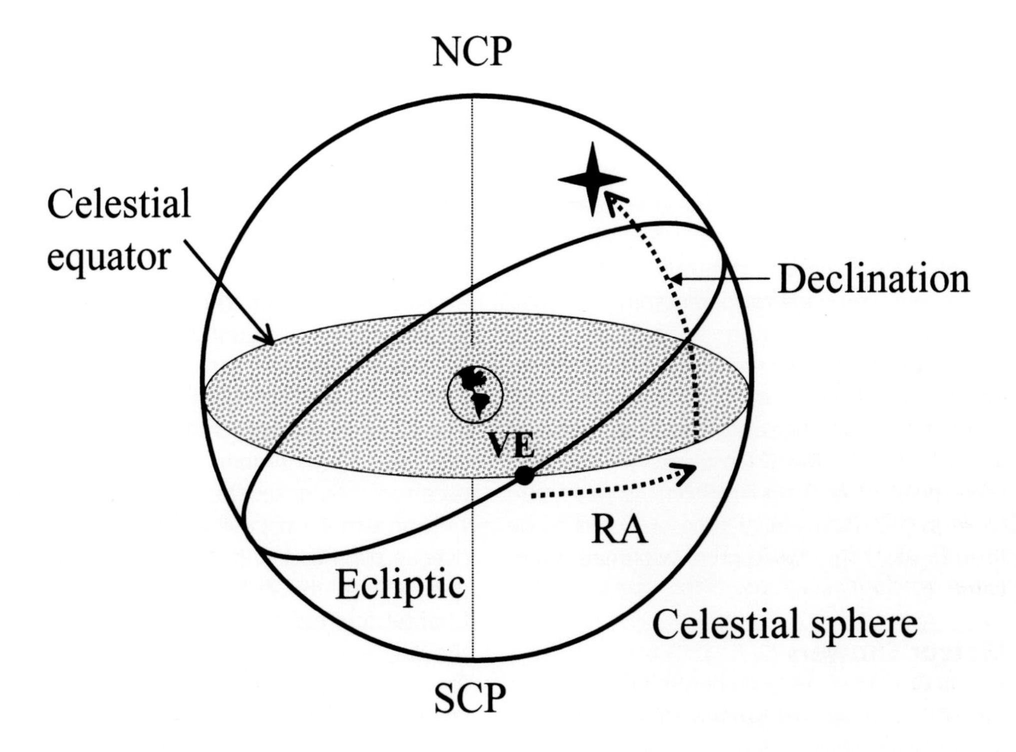

***Fig. 28** The celestial sphere. The line joining the north and south celestial poles (NCP, SCP) is an extension of the Earth's spin axis, and the celestial equator is the projection of the Earth's equator onto the celestial sphere. The ecliptic is the intersection of the plane of the Earth's orbit with the celestial sphere, and is thus the path traced out on the sky by the apparent motion of the Sun during the course of a year. The two points at which the ecliptic and the celestial equator intersect correspond to the vernal or spring equinox (VE) and the autumnal equinox. The right ascension (RA) coordinate is measured in degrees westward along the celestial equator from the vernal equinox. So, for example, RA = 0° corresponds to the vernal equinox, while RA = 180° corresponds to the autumnal equinox. The declination (dec.) coordinate is measured in degrees above or below the celestial equator. Thus dec. = 0° corresponds to a location on the celestial equator, while dec. = +90° corresponds to the north celestial pole. The solar longitude ($\lambda_{\odot}$) of the Earth is also measured in degrees from the vernal equinox, but along the ecliptic.*

calendar, is 12 August 23hr in 2006, 13 August 5hr in 2007, 12 August 11hr in 2008, 12 August 17hr in 2009, and then back to 12 August, 23hr in 2010.

The hour at which the solar longitude reaches 140° in this example is given in Universal Time (UT), which is the time measured from the prime meridian at Greenwich, England. UT is counted from zero starting from local midnight at Greenwich. Astronomers typically express times in UT, but one has to remember that, away from the Greenwich meridian, local time will differ from UT and a time zone correction must be made.

The time zone correction varies systematically with location on the Earth. Observers in central Canada, for example, need to subtract 6 hours from UT to obtain their local time, whereas local time for observers in Western Australia is UT plus 8 hours. Standard time zone maps and tables can be found in almost any atlas. The International Meteor Organization provides a convenient web page for determining the solar longitude corresponding to a given year, month, day and hour: www.imo.net/solarlong/index.html/. There are a number of useful internet sites for checking Universal Time. For example, the US Naval Observatory provides a time display at http://tycho.usno.navy.mil/, while in the UK a similar time service is to be found at http://wwp.greenwichmeantime.com/.

Meteor Showers

A meteor shower occurs when the Earth passes through a meteoroid stream (*see* Figure 29). Since the meteoroids that make up a particular stream are moving along similar orbits about the Sun, when they enter the Earth's atmosphere they will be travelling along nearly parallel paths. They will also have nearly the same initial speeds. Because the meteoroids travel on nearly parallel paths, the meteors they produce will appear to radiate from a small, localized region on the sky. This region is called the radiant, and is discussed in more detail below.

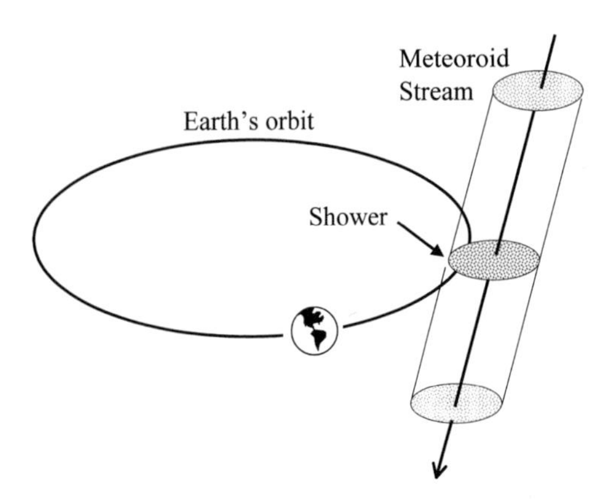

Fig. 29 *When the Earth passes through a meteoroid stream, a meteor shower is seen. The meteoroids that enter the Earth's atmosphere are those that happen to be close to the Earth's orbit when the Earth intercepts the stream.*

Annual meteor showers are so named because they occur during the same period (i.e. around the same solar longitude) each and every year. The activity of a meteor shower is expressed as the **zenithal hourly rate** (ZHR), which is the number of meteors that an observer would see per hour, under perfect viewing conditions, if the shower radiant were directly overhead – at the observer's zenith. (I discuss how the ZHR is actually evaluated in Chapter 5.) The ZHR will vary during a meteor shower, rising from zero to a maximum and then returning to zero again, typically over the course of several days or weeks. Figure 30 shows the activity profile of the Geminids in 2001. The determination of such activity profiles is one of the most important tasks of the modern-day meteor observer.

Meteor showers are described as being major or minor according to the ZHR at the time of maximum activity. In general, if the ZHR is below 10 to 15 at the time of the shower's maximum, then the shower is deemed to be a minor one. Currently the International Meteor Organization (IMO) recognizes 39 night-time annual meteor showers, of which ten are considered to be major showers. The major annual meteor showers for which more than 20 meteors per hour are typically seen at the time of maximum are listed in Table 1. Charts, brief historical notes and observing details for the major meteor showers are provided in the Appendix.

A meteor shower is usually named after the constellation in which its radiant lies on the night of maximum activity. The Taurids, which peak each November, for example, are so

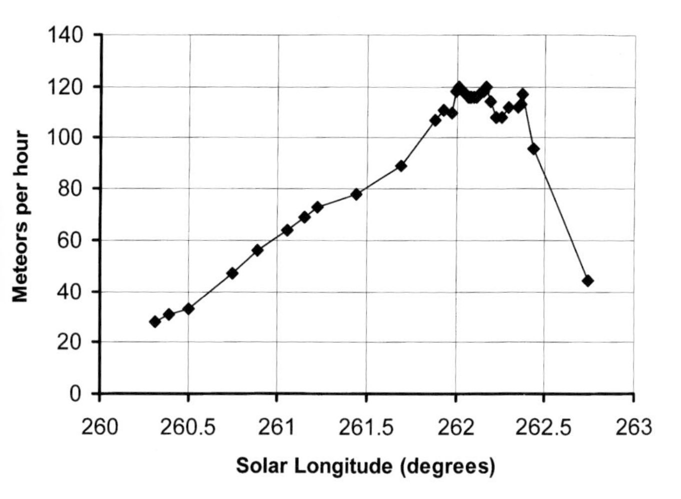

Fig. 30 *Activity profile of the 2001 Geminid meteor shower: zenithal hourly rate (the number of Geminid meteors observed per hour) plotted against the location of the Earth in terms of its solar longitude; the range of solar longitude corresponds to a time interval of about three days. The shower's maximum occurs at* $\lambda_{\odot}$ = *262°, after which there is a rapid decline in the observed meteor rate. (Data from the IMO archive)*

Table 1

Major annual meteor showers. Columns 2 and 3 give the period during which the shower is active and the approximate date upon which maximum activity is seen. Column 4 gives the ZHR at the shower's maximum. Notice that some meteor showers have more than one radiant designation, as is the case for the Northern and Southern Taurids. (Based upon data provided by the International Meteor Organization)

Shower name	Activity limits	Maximum	ZHR
Quadrantids	1–5 Jan.	4 Jan.	120
Lyrids	16–25 Apr.	23 Apr.	20
Eta Aquarids	19 Apr. – 28 May	5 May	50
S. Delta Aquarids	15 Jul. – 28 Aug.	27 July	20
Perseids	17 Jul. – 24 Aug.	12 Aug.	90
Orionids	2 Oct. – 7 Nov.	21 Oct.	20
S. Taurids	15 Sep. – 25 Nov.	5 Nov.	10
N. Taurids	15 Sep. – 25 Nov.	12 Nov.	15
Leonids	14–21 Nov.	17 Nov.	20
Geminids	7–17 Dec.	13 Dec.	120

called because on the night of maximum activity the radiant is in the constellation Taurus. If there is more than one shower with its radiant in a particular constellation, they are distinguished by naming them after a conspicuous star close to the radiant on the night of maximum activity. The Eta Aquarid meteor shower, which occurs in May of each year, is so named because the radiant falls close to the star Eta Aquarii, and distinguishes it from two other, lesser showers, the Delta Aquarids and Iota Aquarids. That said, there are a number of historical foibles associated with some shower names. The Lyrids, for example, are popularly known as the April Lyrids, but should by rights be called the April

Herculids since a constellation boundary change by the International Astronomical Union (IAU) places the radiant, on the night of shower maximum, in the constellation Hercules. And the January Quadrantid meteor shower is named after a constellation which the IAU no longer officially recognizes, Quadrans Muralis.

A number of daylight meteor showers have been discovered by radio and radar techniques (*see* Chapter 7). The IMO presently recognizes twelve daytime meteor showers, of which the strongest are the Arietid and Zeta Perseid showers, which both occur in early June.

A number of meteor showers produce highly variable annual activity profiles. Indeed, some meteor showers have been observed just once. The so-called Corvid shower, for example, was witnessed on just two nights, in June 1937, and has not been recorded since. Other meteor showers, such as the October Draconids, are seen in any great numbers only at irregular intervals. In fact, the Draconids' activity varies in a complex manner related to the return of the stream's parent comet, Giacobini–Zinner, to perihelion. Showers such as the October Draconids and the November Leonids are sometimes called **periodic showers** since their activity varies in step with the orbital period of their parent comets.

Meteor showers tend to be conspicuous for only a few days either side of the time of maximum. Column 2 of Table 1 gives the activity limits for each of the major annual showers. Some showers, such as the Quadrantids, last for just a few days, while others, such as the Perseids, can produce meteors for many weeks. The approximate age of a meteoroid stream can be gauged from its activity profile and its duration. Young, newly forming streams tend to produce short-lived, erratic annual displays. Older meteoroid streams tend to deliver long-lived and consistent annual displays.

Ancient and Modern

Bright fireballs, meteors and meteor showers have been recorded throughout history. The earliest known written account of meteors being observed from a shower that is still active to this day was of the 687BC Lyrids. The Lyrids are not one of the most dynamic of present-day annual meteor showers, but ancient Korean and Chinese court astronomers recorded that during the 687BC display 'meteors fell like rain'. The shower also produced notably strong displays in the eleventh and twelfth centuries AD, and also in 1803 and 1922. Even though they have produced strong displays over many centuries, it was not until 1835 that the Lyrids were recognized as being an annual meteor shower.

Accounts of activity from the Perseid and Leonid meteor showers stretch back a thousand years into the past. The Quadrantids and the Draconids, in contrast, were first recorded in the late 1790s and early 1880s, respectively. One of the youngest known meteor showers is the Pi Puppids, a minor shower visible from the Southern Hemisphere, which is associated with the comet Grigg–Skjellerup. The comet was first seen in 1902, but meteors from an associated shower were not noticed in any numbers until 1977. It would appear, therefore, that the Earth currently 'samples' meteoroids from showers with ages that range from about fifty to about three thousand years. These shower ages, the length of time for which the Earth has been able to sample meteoroids from the various streams, are typically less than the ages of the meteoroid streams themselves.

Neither the parent comet nor its associated meteoroid stream can last for ever, and certainly not for the 4.56 billion years for which the solar system has existed. The active lifetime of a comet is limited by at least three factors. First, once a cometary nucleus has become active it will lose mass each time it passes perihelion. The water ice of which the nucleus is predominantly made will sublimate (i.e. transform straight from a solid ice phase to a gaseous phase) and be lost as gas. It is the outgassing of water vapour that first releases and then drags away the meteoroidal material embedded in the cometary nucleus, thereby producing the distinctive cometary tail and also feeding material into the associated meteoroid stream. There is a finite amount of ice in a cometary nucleus (perhaps around 10^{14}–10^{16}kg) and, as indicated by observed outgassing rates, this is enough for a typical comet to remain active for a few tens of years to perhaps a few hundred thousand years. A second constraint on the lifetime of a cometary nucleus is a close gravitational encounter with a planet, which on average will happen to a comet after about half a million years. During such an encounter a comet's orbit can be radically altered, and in some cases its nucleus can be broken apart. This latter effect appears to be a relatively common occurrence for cometary nuclei, and many comets have been observed to undergo sudden outbursts of activity and shed large fragments from their surface (*see* Figure 31). The most dramatic possible end for a cometary nucleus is to crash into a planet, as comet Shoemaker–Levy 9 did in June 1994, or into the Sun.

Once a meteoroid has been ejected from a cometary nucleus, it can, depending upon its size, survive for some 10^5 to 10^6 years before it is catastrophically destroyed in a collision with another meteoroid. Before then, however, it will more than likely have had its orbit radically altered by planetary encounters. Computer simulations indicate that a meteoroid remains in a stream for perhaps 10^4 to 10^5 years. Compared with the age of the solar system, the meteoroids that produce the meteors we see today are very young indeed*.

In addition, the lifetime of a short-period comet is typically less than that of its associated

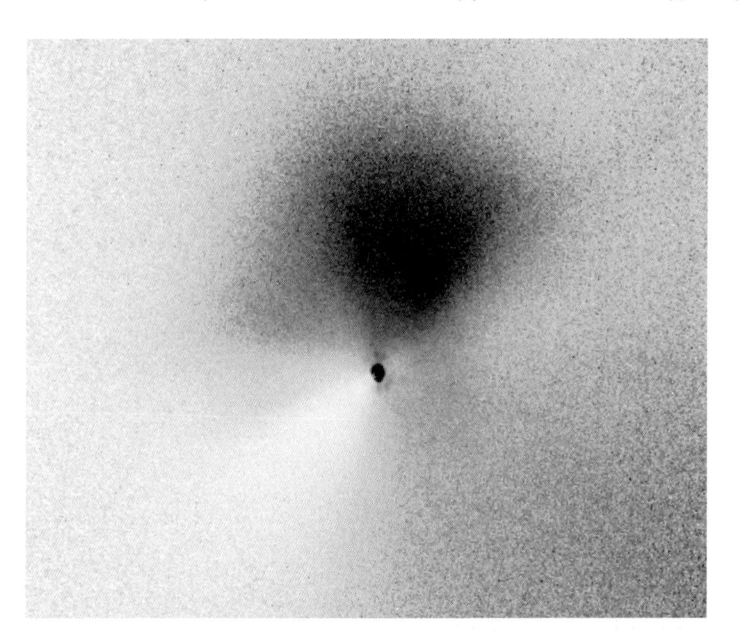

Fig. 31 *An outburst from comet Tempel 1. This Hubble Space Telescope image shows a bright, broad fan-like feature made of newly released dust after the comet underwent an outburst on 14 June 2005. The Deep Impact spacecraft caused the comet to undergo another outburst on 4 July 2005 after it fired a massive copper cylinder into the comet's surface. (Image courtesy NASA/STScI)*

* *By saying that the meteoroids are very young, I refer only to their age since being ejected from a cometary nucleus. The material out of which the meteoroids are composed actually condenced in the early solar nebula sone 4.56 billion years ago.*

Fig. 32 *The multiple fragments of the nucleus of comet Shoemaker–Levy 9 shortly before they crashed into Jupiter. (Image courtesy NASA)*

meteoroid stream, and consequently we should not expect every currently active meteor shower to have a detectable parent comet.

The Radiant

One of the key observational characteristics of a meteor shower is its radiant, the small region on the sky from which the meteors appear to radiate. The radiant is, in fact, an effect of perspective. Just as parallel railway tracks appear to converge at a distant point, so do the parallel paths of shower meteors when traced backwards.

Meteors seen well away from the radiant will in general have longer trails than meteors seen close to the radiant. Importantly, this means that the speed at which a shower meteor appears to move across the sky will vary with distance from the radiant. The angular

Fig. 33 *This image of comet SOHO 6 was captured by the Solar and Heliospheric Observatory (SOHO) spacecraft on 23 December 1996, just before the comet plunged into the Sun. The comet is seen here as a narrow white streak to the lower left of the Sun. (Image courtesy SOHO/ESA/NASA)*

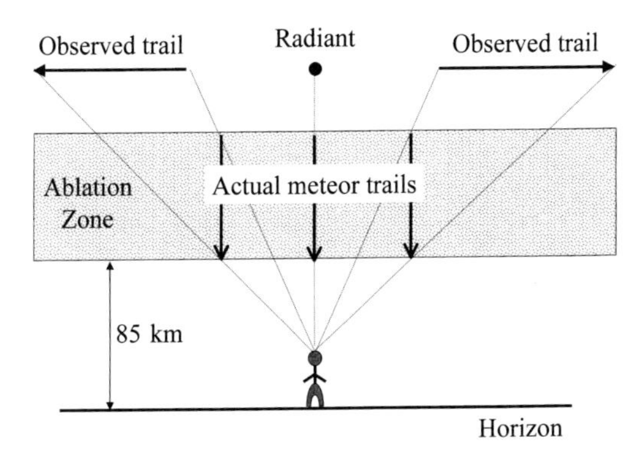

Fig. 34 *A meteor shower's radiant is a perspective effect. Note that meteor trails close to the radiant are shorter than those well away from it.*

speed, which is the number of degrees moved by the meteor across the sky per second, will vary from zero degrees per second for a 'head-on' meteor (i.e. one heading straight towards the observer) to a maximum value when the meteor is nearly 90° from the radiant.

The locations of the radiants of the major meteor showers, in right ascension (RA) and declination (dec.), are given in Table 2. The coordinates are those of the radiant at the time of shower maximum. Figure 28 illustrates the coordinate system of right ascension and declination, which is the standard coordinate system used by astronomers in the construction of star charts. Table 2 indicates the surprising fact that most of the major meteor showers have radiants located in the northern half of the celestial sphere. There is no specific reason for this – it is pure chance. At different epochs in the distant past and in the distant future, the situation was and will be different.

Typically, meteor trails start from a region within 10° of the specified radiant location (i.e. as given in Table 2). The reason that the radiant is not a single point is that the meteoroids in a given stream have slightly different orbital characteristics and speeds: they do not all follow precisely the same orbit at the same speed. In general, the radiant will occupy a larger area of sky for older meteor showers than for younger ones, since the meteoroids in older streams will have accumulated a greater variety of orbital perturbations from

Table 2

The principal night-time meteor showers for which the zenithal hourly rate (column 6) reaches at least 20 meteors per hour on the night of maximum activity. Column 5 gives the initial speed of the meteoroids as they enter the Earth's atmosphere.

Shower	**Solar longitude at maximum**	**Radiant position RA**	**dec.**	**v (km/s)**	**ZHR**
Quadrantids	283.3°	230°	+49°	41	120
Lyrids	32.3°	271°	+34°	48	20
Eta Aquarids	43.5°	336°	–02°	65	50
S. Delta Aquarids	125°	339°	–16°	41	20
Perseids	140.0°	46°	+58°	60	90
Orionid	208°	95°	+16°	66	20
S. Taurids	223°	50°	+14°	27	10
N. Taurids	230°	60°	+23°	29	15
Leonids	235.3°	152°	+22°	71	20
Geminids	262.2°	112°	+33°	35	120

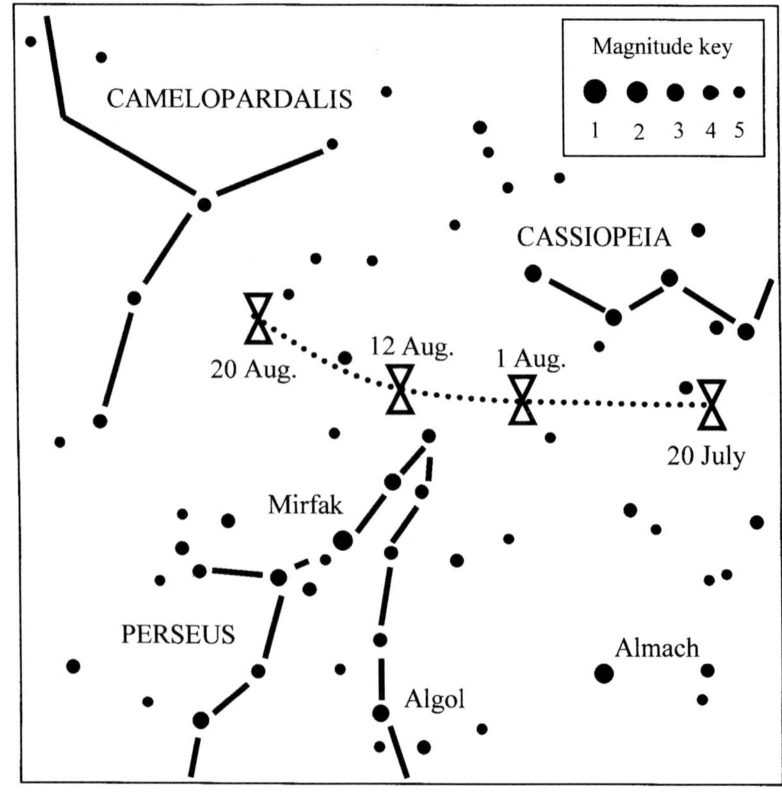

Fig. 35 *The drift (dotted curve) of the radiant of the Perseid meteor shower between 20 July and 20 August. The maximum is on the night of 12 August.*

Table 3

Observed radiant drift parameters for the major annual meteor showers. The drift variations in right ascension and declination are in degrees per day.

Shower	**Daily drift**	
	RA	**dec.**
Quadrantids	+0.4°	–0.2°
Lyrids	+1.1°	0.0°
Eta Aquarids	+0.9°	+0.4°
S. Delta Aquarids	+0.7°	+0.2°
Perseids	+1.3°	+0.1°
Orionid	+0.7°	+0.1°
S. Taurids	+0.8°	+0.2°
N. Taurids	+0.9°	+0.2°
Leonids	+0.7°	–0.4°
Geminids	+1.0°	–0.1°

gravitational interactions with the planets. If we imagine that the 'footprint' of a meteoroid stream, where it cuts through the ecliptic, is a circle, as is illustrated in Figure 29, then the effects of planetary perturbations will be to increase the diameter of the circle with time. A young meteor shower will have a small circular footprint, while an old, evolved shower will have a large one; and the older the shower, the longer it will take the Earth to travel through a shower's footprint.

While the meteors from a given active shower always appear to emanate from a specific radiant, the location of the radiant on the celestial sphere will change during the course of the shower. This effect is most noticeable for older, longer-duration meteor showers such as the Perseids. Table 3 indicates the observed daily drift in the right ascension

(RA) and declination (dec.) for the major meteor showers.

As an example of the use of Table 3, we can consider the motion of the Perseid shower radiant over a 20-day interval centred on the time of the shower's maximum. We need to calculate the radiant's location for solar longitudes between 130° and 150° (the maximum occurs at a solar longitude of 140°, on about 12 August). For times (solar longitudes) before the maximum, we need to subtract appropriate multiples of the drift values from the radiant coordinates given in Table 2. So, at $\lambda_{\odot}$ =130° the Perseid radiant will be located at

RA = 46° – 10 × 1.3° = 33°, dec. = +58° – 10 × 0.1° = +57°

Similarly, for times (solar longitudes) after the maximum we need to add appropriate multiples of the drift values to the radiant coordinates at the time of maximum. At $\lambda_{\odot}$ =150°, therefore, the radiant will be at

RA = 46° + 10 × 1.3° = 59°, dec. = +58° + 10 × 0.1° = +59°

The motion of the radiant on the sky is shown in Figure 35.

The Cometary Connection

The association between comets and meteor showers was established in the second half of the nineteenth century. The key realization at that time was that the orbital elements for several meteoroid streams were almost identical to those that had been deduced for several newly observed comets.

Both comets and meteoroids move about the Sun in elliptical orbits; three cometary orbits are illustrated in Figure 36. The basic shape of an ellipse is described by its **eccentricity**, e, which is between 0 and 1. A circle has zero eccentricity, and the closer e is to 1, the more cigar-shaped is the ellipse. The eccentricity is defined by the ratio e = OF/a (*see* Figure 37), where OF is the distance of the Sun (at F) from the centre of the ellipse (at O), and a is the **semi-major axis** (i.e. half the longest diameter of the ellipse). Since the Sun is not at the centre of the ellipse but at one of the two focal points, the distance between a comet or meteoroid and the Sun as it moves along its orbit will vary from a minimum at perihelion (P) to a maximum at aphelion (A).

The perihelion and aphelion distances are related to the eccentricity and semi-major axis according to the relationships q = a(1 – e) and

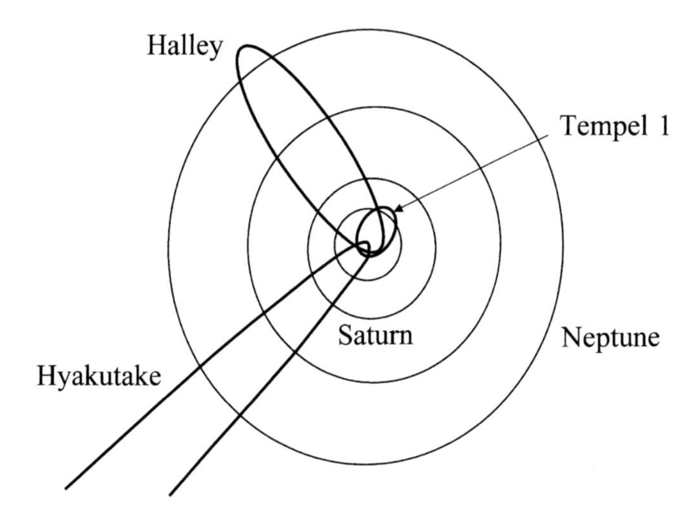

Fig. 36 *The orbits of comets Halley (period 75.4 years), Hyakutake (period 29,500 years) and Tempel 1 (period 5.5 years). Comet Tempel 1 was the target for the Deep Impact spacecraft mission in July 2005.*

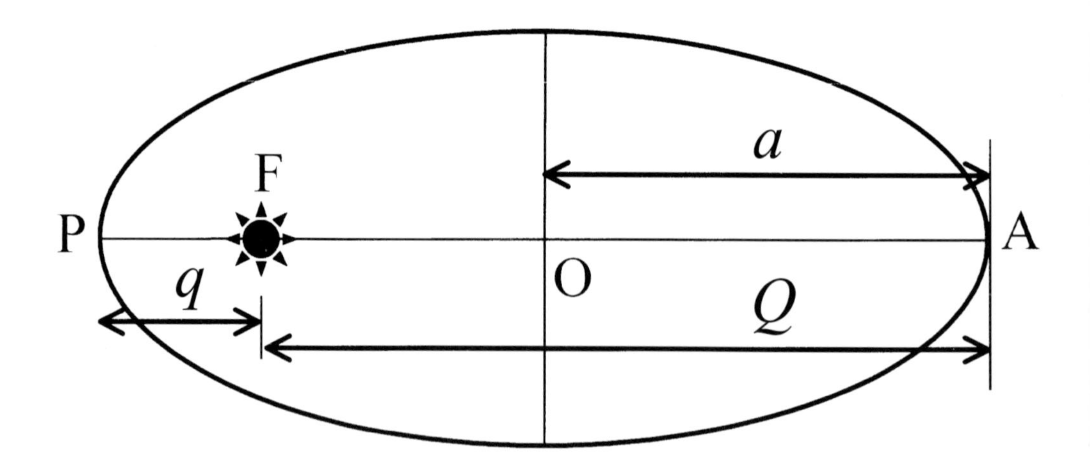

Fig. 37 *Some of the geometrical properties of an elliptical orbit. P is the point of perihelion and A the point of aphelion. O is the centre of the ellipse, and the Sun is at F, one of the ellipse's two focal points. The semi-major axis is denoted by a, and the perihelion and aphelion distances are q and Q, respectively.*

Q = a(1 + e). The orbit of Halley's comet (see Figure 36), for example, has a semi-major axis of a = 17.854AU and an eccentricity of e = 0.9673, and from the above two relationships its perihelion distance is q = 0.584AU and its aphelion distance Q = 35.124AU. At its perihelion point Halley's comet is closer to the Sun than the planet Venus, and at aphelion it is farther from the Sun than the planet Neptune. Kepler's third law provides a correspondence between the orbital period T, measured in years, and the semi-major axis, measured in astronomical units: $T = a^{3/2}$. For Halley's comet, Kepler's third law indicates an orbital period $T = (17.854)^{3/2} = 75.44$ years.

While the eccentricity and semi-major axis describe the basic shape of an elliptical orbit, a further three parameters are required to specify its three-dimensional spatial orientation. These parameters are illustrated in Figure 38. The **inclination** i is the angle between the plane of the orbit and the plane of the ecliptic. The angles corresponding to the **argument of perihelion,** ω, and the **longitude of the ascending node**, Ω, are then used to fully describe the orientation of the orbit in space. By way of example, the orbital elements derived for Halley's comet at its last perihelion passage in 1986 are a = 17.85AU, e = 0.97, i = 162.3°, ω = 111.4° and Ω = 58.4°.

Of major interest to meteor astronomers are the locations of the so-called **nodes**, nA and nD, shown in Figure 38. The subscripts indicate the ascending node (nA) and the descending node (nD), specifying were the comet, or meteoroid, passes from below the plane of the ecliptic to above it, and vice versa. The reason that meteor astronomers are interested in these points is that the Earth can sample only meteoroid streams that have one or both their nodes located close to 1AU from the Sun. A comet's orbit must typically have a node within about 0.1AU of the Earth's orbit before its associated family of meteoroids can produce an annual meteor shower. A few comets produce multiple annual meteor showers. Halley's comet, for example, produces one shower in October, the Orionids, and a second shower in May, the Eta Aquarids. Figure 39 illustrates the geometry that allows a meteoroid stream to produce two annual meteor showers.

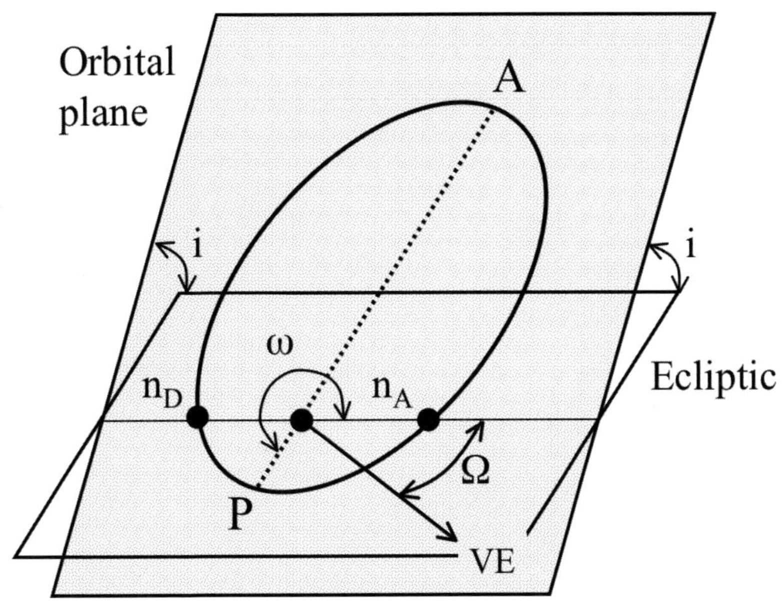

__Fig. 38__ The orbital elements that describe the spatial orientation of an elliptical orbit: the inclination i, the argument of perihelion ω, and the longitude of the ascending node Ω. The vernal equinox is in the direction of VE. P is the perihelion point.

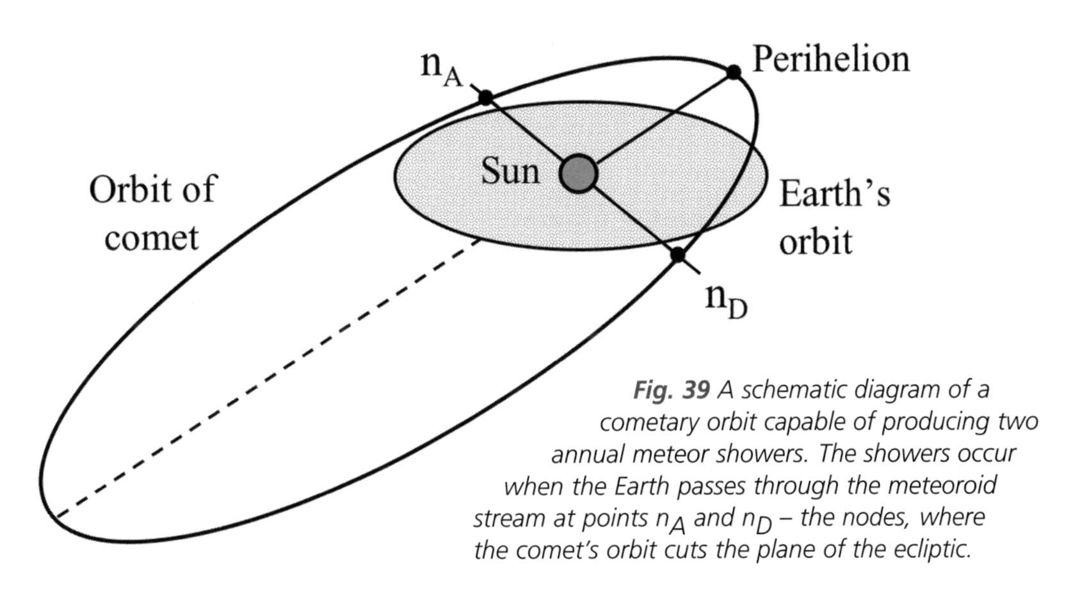

__Fig. 39__ A schematic diagram of a cometary orbit capable of producing two annual meteor showers. The showers occur when the Earth passes through the meteoroid stream at points n_A and n_D – the nodes, where the comet's orbit cuts the plane of the ecliptic.

The orbital elements (a, e, i, ω, Ω) for the Orionid and Eta Aquarid streams are (11.5, 0.951, 164.3°, 82.7°, 28.2°) and (18.3, 0.969, 165.8°, 95.4°, 45.8°), respectively, which compare well with those given above for Halley's comet. The differences in ω and Ω are predominantly the result of gravitational perturbations by the planets.

The Formation of Meteoroid Streams

Meteoroids are ejected from a cometary nucleus each time it rounds the Sun.

Fig. 40 *Halley's comet, photographed on 9 January 1986. The comet shows a distinctive dust tail and coma, and the nucleus is actively adding meteoroids to its associated stream. (Courtesy Lick Observatory)*

Observations indicate that the surface ices begin to sublimate when the comet is within about 2AU of the Sun. Once surface sublimation has commenced, the comet will develop its characteristic coma and tail, and it is then effectively adding meteoroids to its associated stream. The meteoroidal material was originally mixed within the ice at the time the planets formed some 4.56 billion years ago. Since the lifetime of an active short-period comet is typically a few hundred thousand years, the comets we observe at the present epoch must have spent most of their existence since formation well away from the Sun (at least at distances greater than a few astronomical units). It is the action of gravitational perturbations by the planets that cause cometary orbits to change with time, and eventually, for some of the cometary nuclei, the end result is that they pass within a few astronomical units of the Sun and so begin to 'lay down' a meteoroid stream.

Fig. 41 The nucleus of Halley's comet as imaged by the Giotto spacecraft in 1986. The nucleus is some 16 x 8 x 8km in size and composed mostly of water ice, which is sublimating to form the comet's coma and tail. (Courtesy ESA)

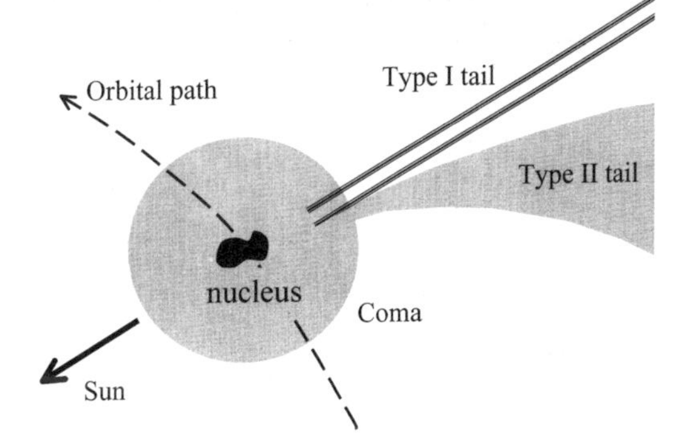

Fig. 42 Anatomy of a comet: nucleus, coma and tails. The cometary nucleus is typically just a few kilometres to a few tens of kilometres in size, whereas the coma can be hundreds of thousands of kilometres across. It is sunlight reflected from the dust within the coma that is observed when we see a comet from the Earth. The straight, type I tail is produced by ionized particles interacting with the Sun's interplanetary magnetic field. The curved, type II tail is composed of very small dust particles (and meteoroids). Both tails can extend away from the nucleus to distances of millions of kilometres.

The meteoroids are carried away from the cometary nucleus by the gas outflow produced by the sublimating ice. Meteoroids leave the cometary nucleus at speeds of typically a few hundred metres per second. These speeds are additional to the much higher speed the meteoroids have by virtue of the comet's orbital motion about the Sun. Since the meteoroids are ejected in random directions, some of them will have slightly higher speeds than the parent nucleus, while others will have speeds that are slightly lower. This variation in relative velocity means that some meteoroids will acquire orbits that subsequently bring them back to the Sun before their parent comet, while others will arrive at perihelion later than the comet. In this manner the meteoroids gradually, after many years and many orbits of the Sun, begin to spread away from the comet. Figure 44 shows

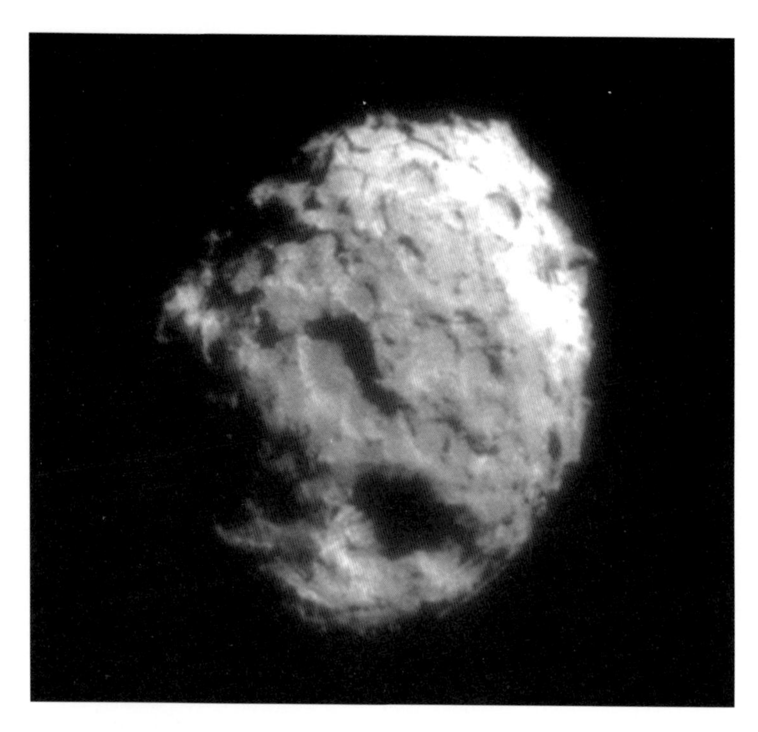

Fig. 43 *The nucleus of comet Wild 2 as imaged by the Stardust spacecraft on 2 January 2004. The nucleus is about 5km in diameter and is seen from a range of about 500km in this image. Most of the comet's surface is not active, it having developed a thick insulating mantle. The outgassing is restricted to active regions that cover just a few per cent of the comet's total surface area. (Image courtesy NASA/JPL)*

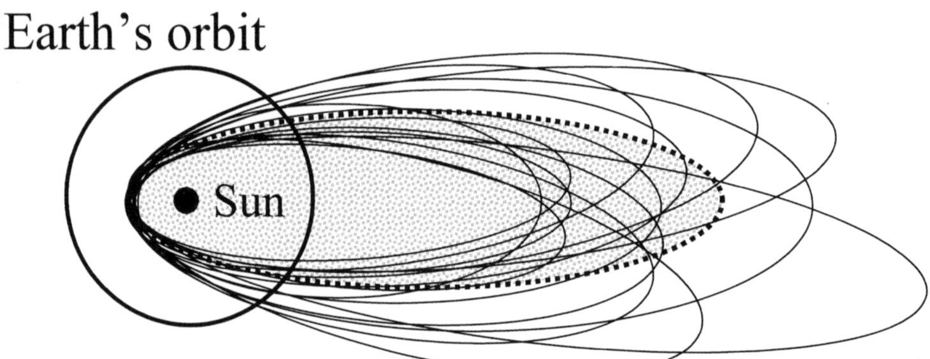

Fig. 44 *A schematic drawing of a meteoroid stream. Some meteoroids have orbits that are smaller and some very much larger than the orbit of the parent comet (shown by the dotted line and shaded ellipse).*

schematically the range of orbits that might be found in a typical meteoroid stream.

Once individual meteoroids are ejected from a comet, their orbits are disturbed by the Sun's radiation and by the gravitational attractions of the planets. It is these accumulated perturbations that eventually cause the meteoroids within a stream to adopt orbits increasingly different from that of the parent comet. If a meteor shower is to be seen at some stage, then at least some of the meteoroids must acquire orbits that bring them close to the Earth. In most cases only a very small fraction of the meteoroids ejected from a

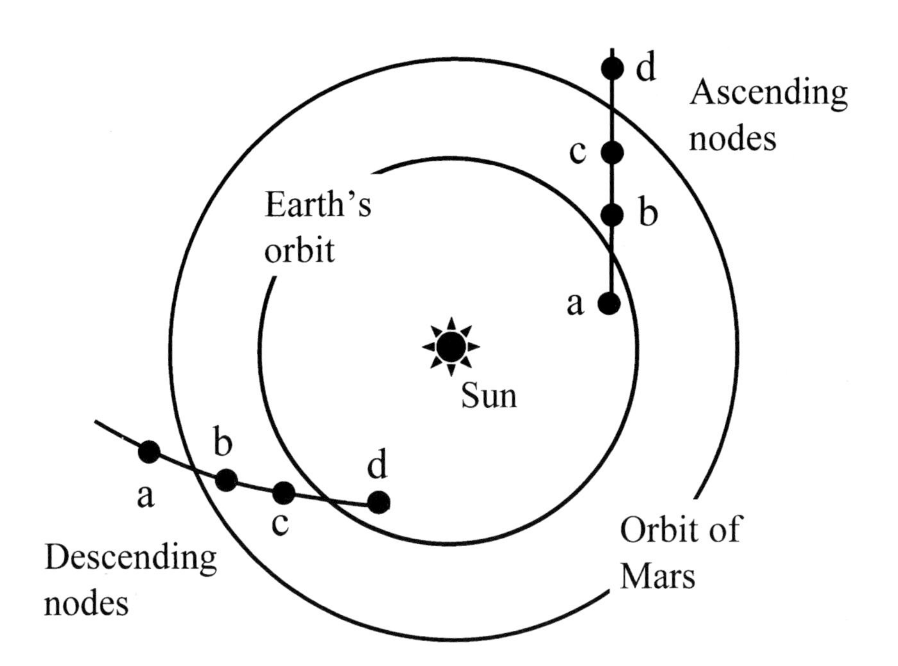

Fig. 45 *A schematic diagram of the two ecliptic-crossing arcs for meteoroids ejected from Halley's comet during the past 3,500 years. The letters a, b, c and d indicate the location of the nodal points in 1400BC, 700BC, AD70 and 1986, respectively. The Earth samples only a very small part of the stream's cross-section. Note that Halley's comet also produces a Martian meteor shower.*

comet will acquire orbits that allow them to be sampled by the Earth. We literally see the tip of the iceberg when we observe meteor showers – the vast majority of the meteoroids in any stream will sail past the Earth, never to produce a meteor in our sky. Figure 45 shows the extent of the region over which meteoroids ejected from Halley's comet during the past 3,500 years cut through the plane of the ecliptic.

The Asteroid Connection

While all the major annual meteor showers with known parent bodies are clearly associated with comets, the situation is less clear for some of the minor streams. Indeed, during the past few decades a number of researchers have argued that some asteroids may have associated meteoroid streams. Here, 'associated' means that the orbital elements of a particular asteroid are similar to those deduced for a meteor shower. It has been suggested, for example, that the daytime Arietid meteor shower is associated with the asteroid Icarus. It seems, therefore, that some objects which have been designated as asteroids are actually old and no longer active cometary nuclei. This is the case with the Geminid meteoroid stream, for which the parent object Phaethon has an asteroidal rather than a cometary designation. The observed characteristics of Geminid meteors, however, are very different from those of meteorite-producing fireballs, and the cometary nature of Phaethon is thus betrayed. Some researchers have suggested that meteorite-producing streams do exist, but the evidence for such streams is far from conclusive and there is no consensus on the matter.

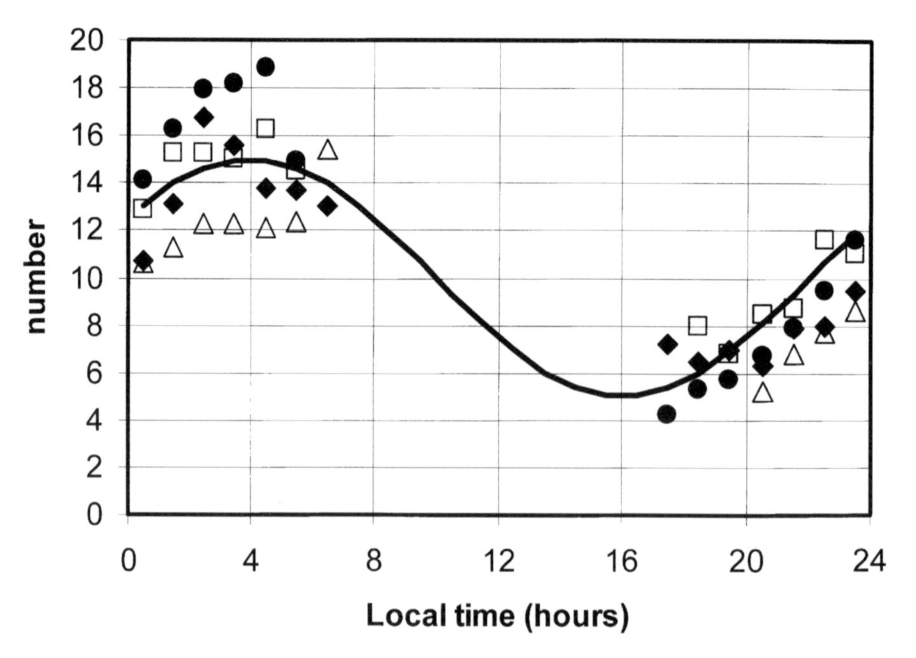

The Sporadic Background

The diurnal (daily) variation in the number of visual sporadic meteors is shown in Figure 46, from which it can be seen that there is a systematic variation in the number of sporadic meteors visible per hour over the course of a day. The sporadic meteor rate reaches a maximum at around 4am local time and falls to a minimum at about 4pm local time. The sporadic rate at its minimum is about one-third of the maximum rate.

The annual variation in the daily averaged sporadic meteor rate is shown in Figure 47, from which it can be seen that greater numbers of sporadic meteors are recorded during the latter part of the year. Sporadic meteor rates are at their highest at about the time of the autumnal equinox (22 September), and at their lowest near the spring equinox (20 March). There is a variation of about a factor of two and a half between the autumnal and spring equinox daily average sporadic meteor rates.

Why the sporadic meteor rate should vary over the course of the year is presently not fully understood. Certainly, however, some of the

***Fig. 46** The diurnal variation in the number of visual sporadic meteors. The data points indicate actual sporadic meteor rates (averaged over many years) from four experienced visual meteor observers. (Data from T. Murakami,* Publications of the Astronomical Society of Japan, *vol. 7, p. 49, 1955)*

annual variation relates to the continuously changing altitude in the observer's sky of the apex of the Earth's way – the instantaneous direction in which the Earth is moving in its orbit about the Sun. The apex is always 90° west of the Sun and located on the ecliptic. The apex transits at 6am local time, when an observer looking due south will be facing the direction in which the Earth is moving. The altitude of the apex, however, will vary according to the time of year and the observer's latitude on the Earth.

In contrast to the yearly variation, the reason for the variation in the sporadic meteor rate over the course of a day is reasonably clear. Once again, the reason relates to the location

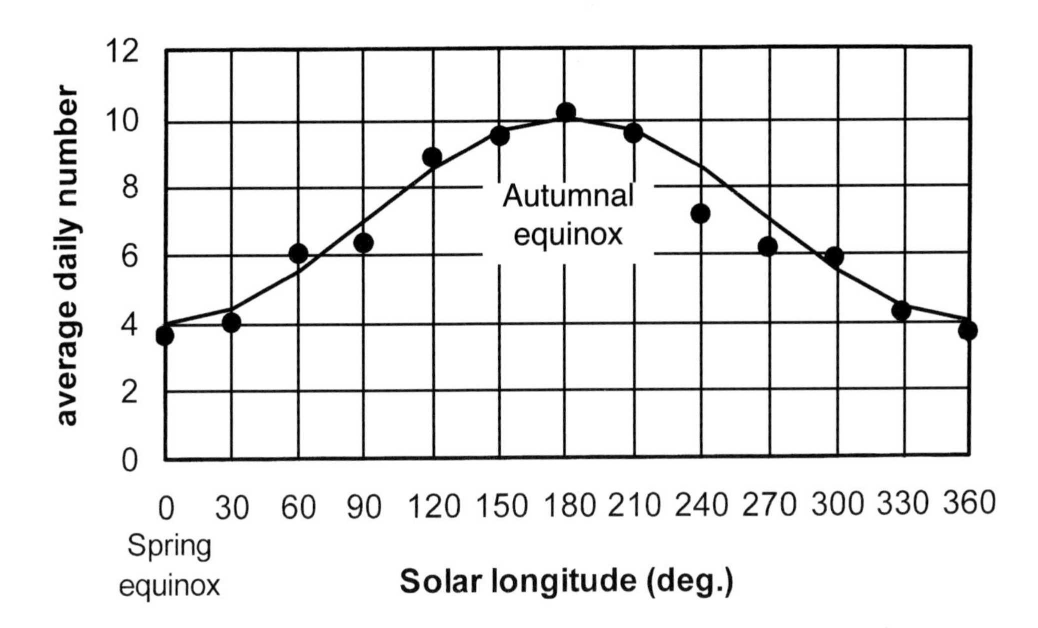

Fig. 47 *The annual variation in the daily averaged number of visual sporadic meteors in the northern hemisphere. The approximate reverse of this diagram holds for observers in the southern hemisphere. (Data points are estimates taken from T. Murakami,* Publications of the Astronomical Society of Japan, *vol. 7, p. 49, 1955)*

of the observer with respect to the apex of the Earth's way. As shown in Figure 48, at 6pm local time an observer will be facing directly away from the direction of the apex, so any meteoroids capable of producing meteors visible to the observer will be approaching the Earth from behind, and will have to catch up with it, and the relative velocity of the meteoroid and the Earth will be at its lowest possible value. The exact reverse of this situation holds true at 6am local time: now the Earth will intercept meteoroids in a head-on collision, and consequently the relative speed will be at its greatest possible value.

At this stage all I have pointed out is that the typical relative encounter speed between the Earth and a meteoroid will vary according to whether the encounter is near 6am local time (head-on collision) or 6pm local time (catch-up collision). Why does this translate into more meteors being seen? Essentially, the meteors seen in the morning hours will be faster and brighter than those seen in the early evening. Now, recall from equation (1) (page 19) that a meteor's brightness increases with increasing velocity. Consequently, a smaller, less massive meteoroid encountered in the early morning will, because of its higher velocity, achieve the same visual brightness as a larger, more massive meteoroid, moving more slowly, encountered in the early evening. And because there are more small-mass meteoroids than large-mass ones, more meteors brighter than a given limiting magnitude will be seen in the early morning hours.

While more sporadic meteors are seen in the morning than in the evening because of the orientation of the observer with respect to the direction of the apex of the Earth's way, the exact opposite applies to the observed number of meteorite falls. For a meteoroid to survive its

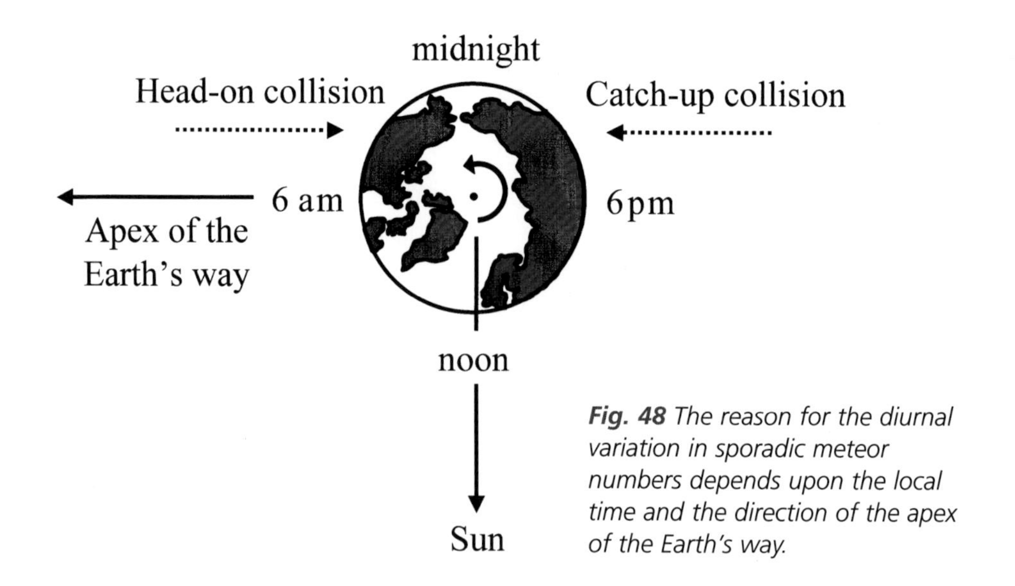

__Fig. 48__ The reason for the diurnal variation in sporadic meteor numbers depends upon the local time and the direction of the apex of the Earth's way.

passage through the Earth's atmosphere to become a meteorite, its initial speed must, in general, be less than about 20km/s. At higher speeds, so much mass is lost through ablation that nothing survives to hit the ground. A low encounter speed is more likely if a meteorite-producing meteoroid enters the Earth's atmosphere in a catch-up manner – that is, at a local time close to 6pm. And indeed, the detailed study of meteorite fall times reveals that most meteorites do fall during the afternoon at local times between 4 and 6pm.

3 Observing Fireballs

Meteor fireballs are a magnificent sight. When one appears, the whole sky seems to tremble with expectation and excitement. They are awe-inspiring and they are sublime. Not only can they turn, for a few brief seconds, dark night into brilliant day, but they can sometimes rumble across the heavens, producing thunderous booms and crackling sounds. Not every fireball heralds a meteorite falling to the ground, but a fireball certainly presages every meteorite.

Fireball Reporting

'Fireball' is the name reserved for the brightest of meteors, and by formal definition it is used for any meteor that has a maximum brightness greater than that of the planet Venus, the brightest of the planets. This brightness definition requires that a meteor should be brighter than about magnitude –4.7 before it can be called a fireball. Perhaps only one out of several hundred visible meteors will become brighter than magnitude –4.7, while only one in ten thousand will be brighter than

Fig. 49 *The fireball associated with the Peekskill meteorite. This image, captured by S. Eichmiller (of the* Altoona Mirror *newspaper) was taken on 9 October 1992 from Altoona, Pennsylvania. The photograph shows a multitude of fireball components, which indicates that the meteorite had broken into numerous fragments. The photograph was taken with a 300mm, f/2.8 lens with an exposure time of 1/500 of a second on 3200 TMAX black and white film.*

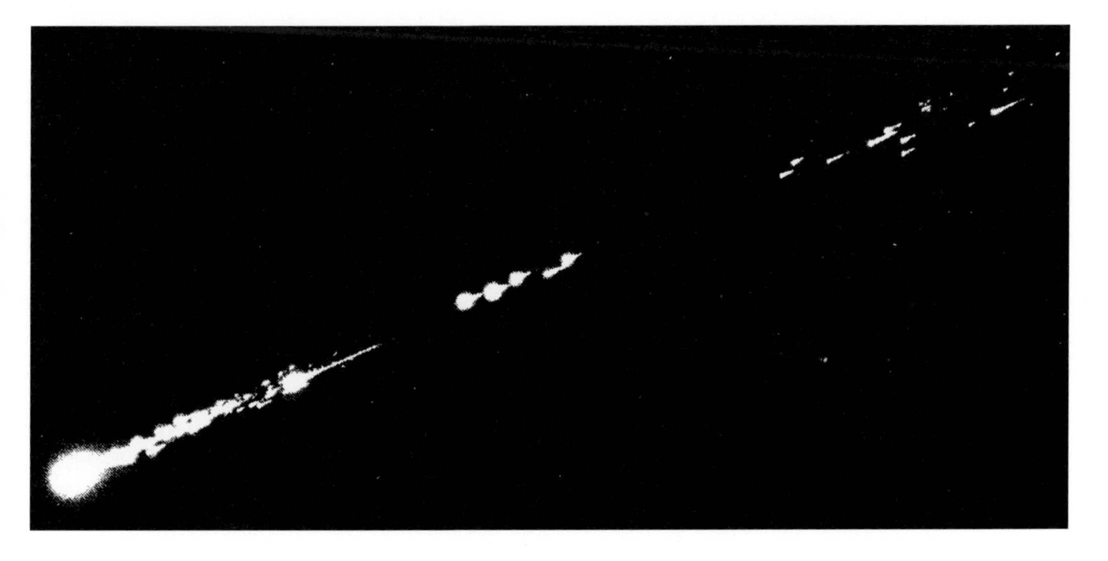

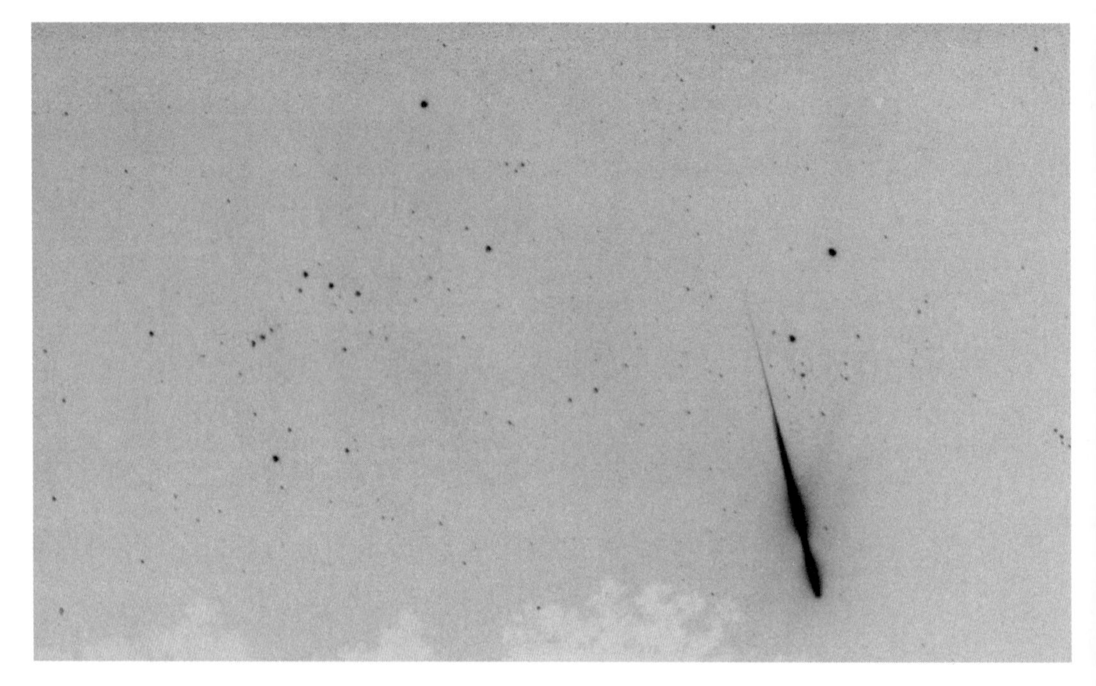

Fig. 50 *A brilliant Leonid fireball is recorded in this image against the backdrop of the constellation Taurus. (Original image by G. Holmes, http://leonids.hq.nasa.gov/leonids/)*

magnitude –8. Fireballs as bright as the full Moon (magnitude –12.7) are seen perhaps only a few times in a dedicated observer's lifetime.

If you are fortunate enough to see a spectacular fireball while out observing other meteors, there are a number of details that you should try to record. The main points to note are the following:

- Your location. Make a note of exactly where you were, and also the direction in which you were facing – it is amazing how quickly such details become 'fuzzy'.
- The date and time of the event in Universal Time. An accurate UT time can be obtained from a portable Global Positioning System (GPS) receiver – such receivers will also provide an accurate latitude and longitude for your location.
- The apparent path of the fireball on the sky. Record it on a star chart, or if you don't have one to hand, try to note which bright stars the fireball passed close to. Draw a sketch of the fireball path as soon as you can after the event. If you have a compass with you, record the azimuth and altitude of the first and last positions on the sky where you saw the fireball.
- An estimate of the maximum brightness. This can be very difficult for fireballs since there are often no astronomical objects of comparable brightness. If the Moon, Venus or Jupiter is in the sky, then use it as a guide to the magnitude.
- An estimate of the duration of the fireball in seconds.
- Any coloration in the trail. Do this once you have the main characteristics of the fireball recorded. Some fireballs, for example, display a distinct green, blue or red colour.
- Any sudden changes in brightness. Flares (brightness enhancements lasting just a fraction of a second) are often seen, and so too are terminal detonations, when it seems

that the fireball has exploded. In reality it has catastrophically broken apart into many very small fragments that are rapidly destroyed in the Earth's upper atmosphere. On other occasions, more gradual fragmentation will occur, resulting in the appearance of multiple trails (see Figure 49). If that happens, try to record how many fragments are present and where in the trail they occurred – that is, near the beginning or the end of the path as you saw it.

- Whether there is a luminous train. Bright fireballs often leave such trains of luminescence, which can sometimes last for many minutes, in their path. The origin and physics of such trains are only poorly understood at the present time, and any observations (especially by photography or video) will be of scientific value. If you don't have a camera to hand, then try to draw a series of diagrams showing the shape and sky location of the train at regular time intervals.
- What sounds were heard. Sounds are an important indicator of how deeply a fireball has penetrated into the Earth's atmosphere, and of whether material may have survived to reach the ground, as meteorites. Various audible phenomena have been associated with the passage of fireballs, and are discussed in greater detail below. What is most important to record is when the sounds were heard: whether before, during or after the fireball was seen. If the sounds were heard after the fireball became extinct, you should try to estimate the time delay between the time of extinction and when sounds were first heard.

Since neither the time nor the path of a fireball can be predicted, most eyewitnesses are caught completely off-guard when such an event lights up the heavens (bright fireballs can even be seen during daylight hours). This invariably means that most of the important information, such as the start and end points of the trail, is only poorly determined. The key, therefore, to investigating fireball events is to act immediately – even a single day's delay can lead to the loss both of eyewitnesses and of the acquisition of useful data. Although the basic characteristics and direction of a fireball can often be determined by collecting many eyewitness accounts, the results thus obtained are generally not very satisfying and are typically highly uncertain. By far the best way to determine a fireball's atmospheric path is to set up a network of all-sky camera systems. Such systems, if correctly calibrated, can not only record the atmospheric path of the fireball but also reveal the parent meteoroid's orbit before it encountered the Earth. Such photographic records are particularly useful

Fig. 51 *An all-sky camera located at Streitheim in Germany. The camera is housed at the top of the tripod structure and looks down upon the spherical mirror. (Image courtesy D. Heinlein)*

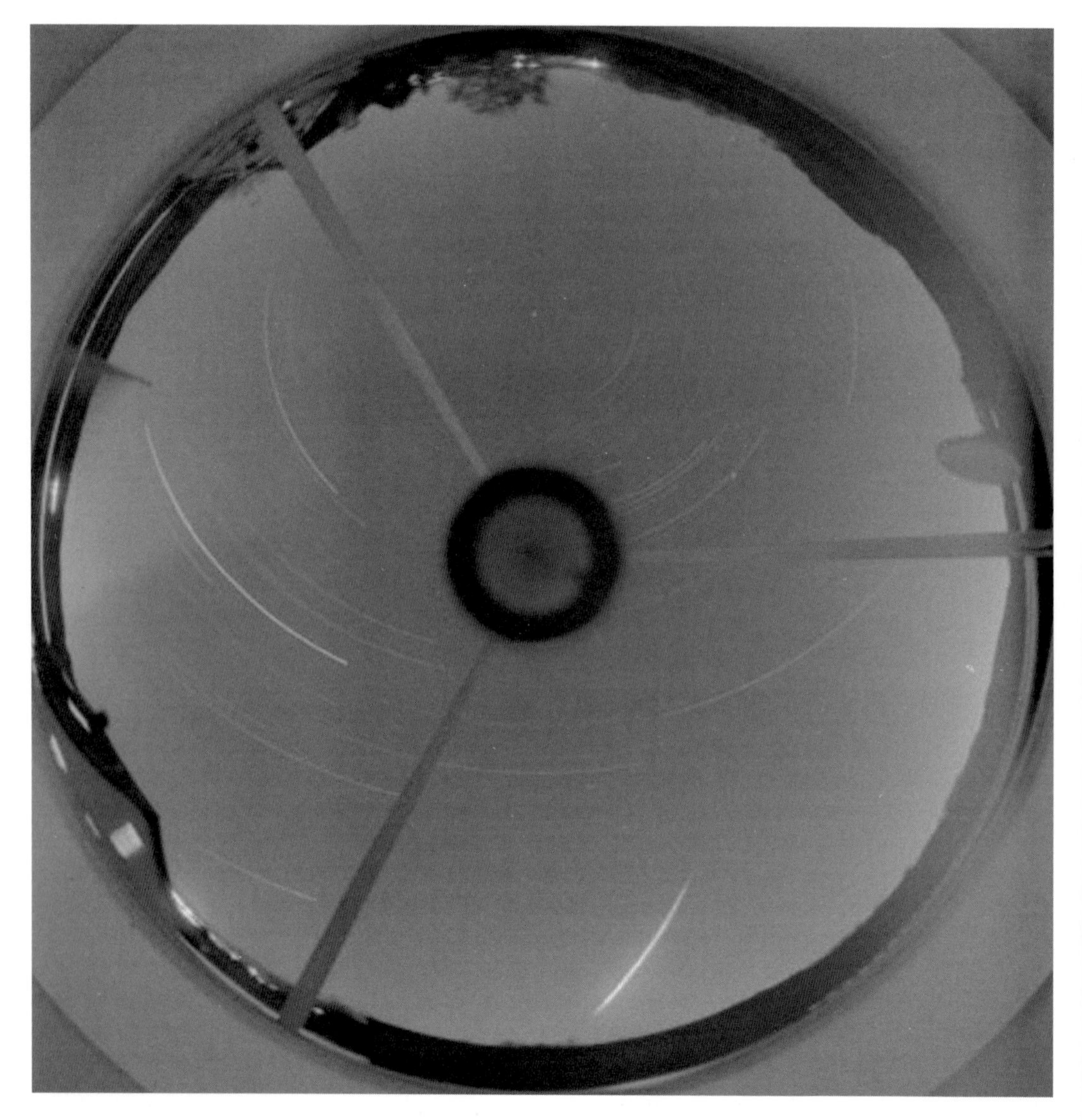

Fig. 52 *The fireball that heralded the arrival of the Neuschwanstein meteorite. The central circle in the image is the camera housing, and the outer circle shows the 360° sweep of the horizon. The fireball is the bright streak to the lower right. The fireball first became visible at an altitude of 85km, when it had a speed of 21km/s. Its maximum brightness, magnitude –17, was recorded during a bright flare at a height of 21km. (Image courtesy D. Heinlein)*

when it comes to possible meteorite falls since they enable accurate ground impact locations to be determined, often to within a few hundred metres or so. Figure 51 shows one of the all-sky camera systems in the European Network of fireball cameras. This particular camera recorded the fireball (*see* Figure 52) associated with the fall of the Neuschwanstein meteorite on 6 April 2002.

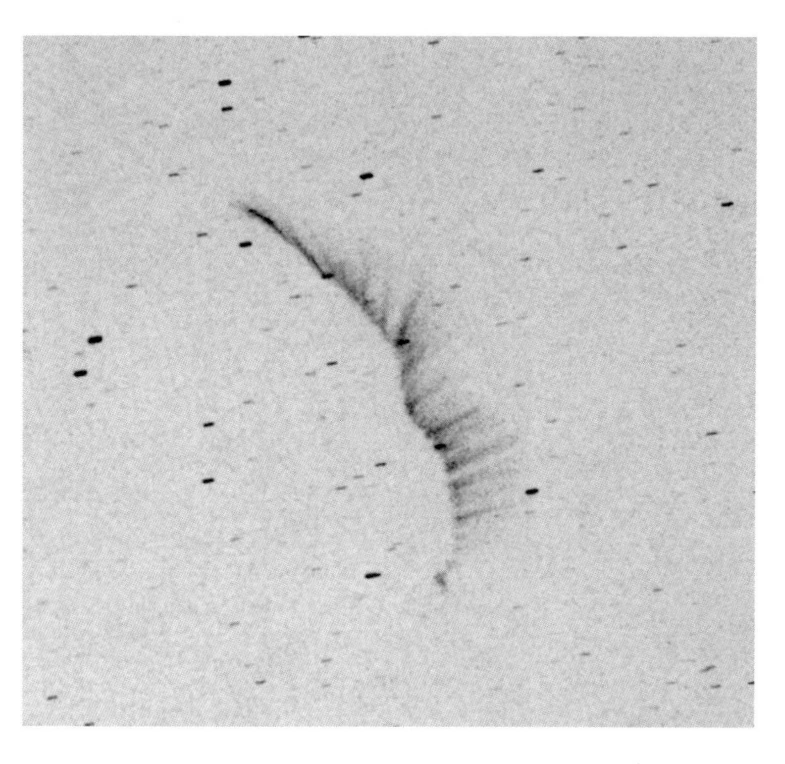

***Fig. 53** A distinctive persistent train produced by a magnitude –6 Leonid fireball. The train was distorted by high-altitude winds and also drifted slightly during the 2-minute exposure. (Courtesy J. W. Young, TMO/JPL/NASA)*

Persistent Trains

One of the rarest and least well understood phenomena associated with fireballs is the occurrence of a persistent train – a faint but long-lived greenish glow that appears in the path of the fireball. Such trains can last from a few seconds to, on very rare occasions, an hour or more. The brightness of a train is invariably much less than that of the fireball that produced it, and, unlike the straight path that the fireball carves across the sky, a train will gradually twist and turn into a random, 'tortured' shape (*see* Figure 53). The twisting of the train is caused by different wind speeds at different altitudes in the Earth's atmosphere.

Observations indicate that large and/or fast meteoroids typically produce persistent trains. Indeed, short-duration trains are often observed to accompany the high-velocity meteoroids associated with the annual Perseid and Leonid showers. Persistent trains occur at heights between 70 and 110km. Figure 54 shows a series of four images taken by the Robotic Optical Transient Search Experiment (ROTSE) camera system located at Los Alamos, New Mexico. The images reveal the developing distortion of a bright Leonid train. The beginning train height was determined to be about 83km, and the train remained visible for 34 minutes.

Smoke Trails

A second form of trail, often seen after meteorite-dropping fireball events, is not self-luminous but shines by reflected sunlight (*see* Figure 55). These trails are composed of small dust grains, the remnants of ablation products left in the track of the meteoroid as it descends through the Earth's atmosphere, and are known as smoke trails. Since smoke trails require sunlight to be seen, photographing them is a relatively straightforward exercise, and ordinary daylight film has been used successfully. Ordinary video cameras have also

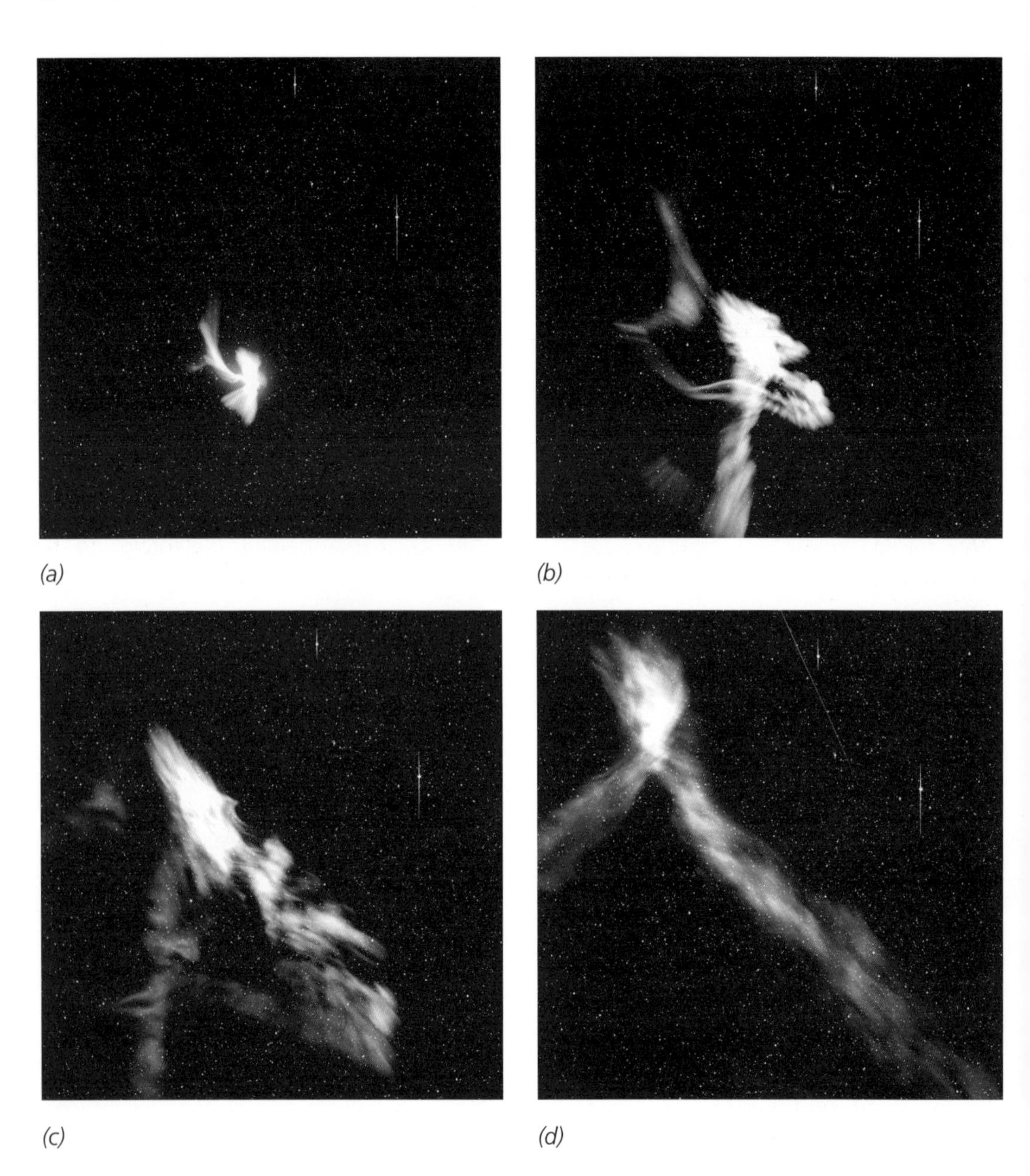

(a) (b) (c) (d)

Fig. 54 *Images of a Leonid fireball train captured on the night of 17 November 1998. Each image is a 60-second exposure, commencing at (a) 1:32:33, (b) 1:35:36, (c) 1:42:42 and (d) 1:52:52 (all times UT). (Images courtesy the ROTSE project, http://rotse1.physics.lsa.umich.edu/)*

been successfully used to record how smoke trails change over time.

If you chance to see a smoke trail and have a video camera, then record it with the camera attached to a tripod (or at least a solid support) to minimize camera shake. Most importantly, try to capture some foreground objects in the

Fig. 55 *The smoke trail left after the fall of the Tagish Lake meteorite, on 18 January 2002. Ewald Lemke of Atlin, British Columbia, took this photograph shortly after the fireball had been seen to streak across the sky. High-altitude winds distorted and fragmented the originally straight trail. (Image courtesy E. Lemke)*

image – these can be very useful for fixing the trail's altitude and azimuth, because you can take compass bearings of the foreground objects at a later time, once the video/photographic images have been successfully captured. Also, do not use the zoom feature – or if you do, then use it very sparingly – for every time you adjust the camera zoom the image scale will change, and useful information on the trail's spatial motion will be lost.

Three videographers serendipitously captured the daylight fireball associated with the Morávka meteorite fall in 2000. Figure 56 shows a series of four individual frames from the video recorded by J. Fabig from Janov in the Czech Republic. The actual video sequence lasted for about a second before the fireball passed behind the foreground building, but this video and the two others did enable the fireball's atmospheric trajectory, and hence the orbit of the progenitor meteoroid, to be calculated.

Sonic Booms and Simultaneous Sounds

Imagine the scene: a brilliant fireball has just streaked across the darkened sky. For a few heart-stopping seconds, night has turned into day. The observer is left breathless and awestruck. And then, perhaps a few minutes after the fireball has passed, just when the world seems to be getting back to normal, a

Fig. 56 *Four individual frames from the video sequence of the Morávka meteorite fireball on 6 May 2000. Because the exact location of the videographer was known, the position of the fireball in the sky could be determined relative to the measured positions of features of the building. (Images courtesy J. Fabig and J. Borovicka)*

low rumbling is heard in the distance. Clashing and rolling, the sound builds to a crescendo. The observer has just heard the sonic booms associated with the passage of a large meteoroid through the Earth's lower atmosphere.

Over the years, a number of schemes have been proposed for the classification of sounds associated with fireballs. The most straightforward of them simply divides sounds into two categories according to when they are heard, either 'sonic' or 'simultaneous'. The former are characterized by the fact that they are heard several minutes after the fireball has passed, while the latter are anomalous in that they are heard at the same time as the fireball is seen in the sky.

Sonic booms are caused by shock waves generated in the Earth's lower atmosphere. Essentially, a fireball produces a cylindrical shock wave as it descends, at hypersonic speed, through the Earth's atmosphere. The propagation of the shock wave and the location of the audibility zones, where the sonic booms can actually be heard, are determined by local atmospheric conditions and prevailing winds. In some cases one observer will hear nothing, while a second observer, just a few kilometres away from the first, will hear the sonic booms in all their glory.

Simultaneous sounds are distinguished from sonic booms in that they are heard when the fireball is seen, not later. They are typically described as 'crackling', 'hissing' or 'screeching', in contrast to the rolling, thunder-like sonic booms. On the face of it, from a straightforward physics perspective simultaneous sounds appear to be impossible. At sea level, for example, sound propagates at a speed of 330m/s; higher in the atmosphere it propagates more slowly. If a fireball is witnessed at a range of, say, 40km, it should take about two minutes for the sound waves to reach the observer. Clearly, this doesn't square with the observation that sounds are heard simultaneously with seeing the fireball.

In the early 1980s, Colin Keay of the University of Newcastle, Australia, suggested that the simultaneous generation of sound could be explained by an interaction between the fireball's ionized tail and the Earth's magnetic field. In Keay's model, often described as the 'magnetic spaghetti model', there are two main interactions (see Figure 57).

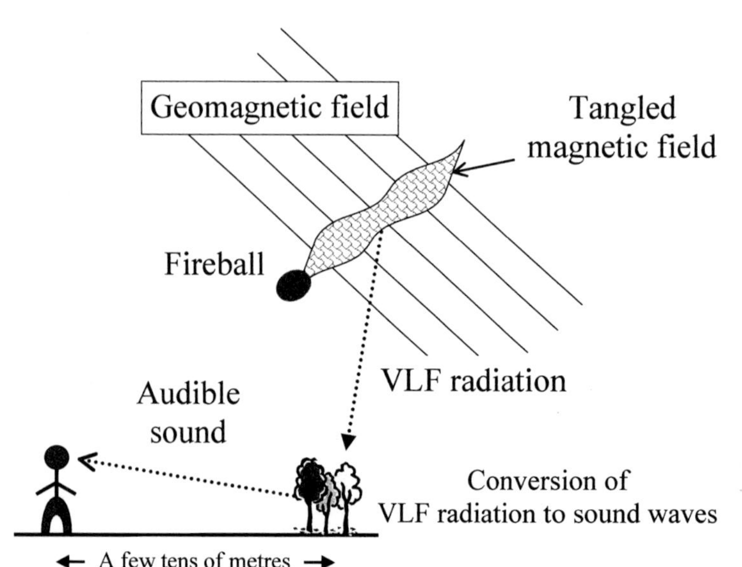

***Fig. 57** A schematic depiction of the 'magnetic spaghetti model' for generating simultaneous sounds. The twisting of geomagnetic field lines by the fireball causes radiation of very low frequency to be emitted. The actual sounds are produced by the conversion of the VLF radiation into audible sound by trees or other vegetation close to the observer.*

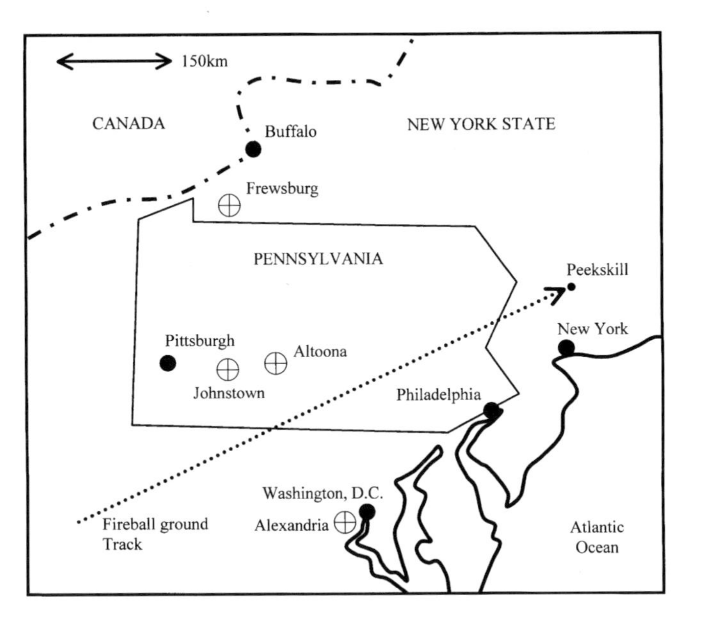

Fig. 58 *The ground track of the Peekskill fireball. The open circles with a cross show the locations from which simultaneous sounds were reported. At Johnstown a 'rushing' sound was reported. A crackling sound lasting about 10 seconds was reported by several eyewitnesses at Altoona. Low, 'whooshing' sounds were reported from Frewsbury, while in Alexandria a horse and rider were 'spooked' by electrophonic sounds. Details of the sound reports can be found in the journal* Earth, Moon and Planets, *vol. 68, p. 189, 1995.*

First, the twisting of the Earth's magnetic field lines by the fireball's tail produces very-low-frequency (VLF) radio waves. The frequencies of these radio waves is well outside the range of standard, commercial radio receivers, and are typically just a few kilohertz. The second step in Keay's model requires the VLF radio waves to interact with objects, such as trees or other vegetation, close to the observer. These objects convert the VLF radiation into higher-frequency sound waves that are audible to the observer. Since radio waves travel at the speed of light, and the conversion takes place at the observer's location, there is no problem with the sound being heard at the same time as the fireball is witnessed.

Because Keay's model invokes the generation of VLF radio waves, simultaneous sounds are often called **electrophonic sounds**. One of the most intriguing aspects of simultaneous sounds is that they are highly localized. Two observers just a few hundred metres apart, for example, may sense very different sounds, or – as is common – one observer will hear a distinct sound and the other will hear absolutely nothing. Keay has found some evidence to suggest that having long, frizzy hair and wearing spectacles can make simultaneous sounds easier to hear. Being near trees or to sparse, dry ground vegetation also helps.

Simultaneous sounds are reported to have been heard at distances of several hundred kilometres from a fireball's ground path. In the case of the fireball associated with the fall of the Morávka meteorite (*see* Figure 56), electrophonic sounds were heard as far as 250km away from the fireball (sonic booms were reported from distances up to 100km away). No sonic booms were reported for the fireball associated with the fall of the Peekskill meteorite (*see* Figure 49), but electrophonic sounds were reported from as far as 250km from the ground path.

Sounds are not heard from every fireball, but it is well established that all fireball events that result in meteorites landing on the ground generate sounds, both sonic and simultaneous. (On this basis, many meteorite searchers will use the locations of observers who heard sounds to narrow down their search area.) However, not all sound-generating fireballs

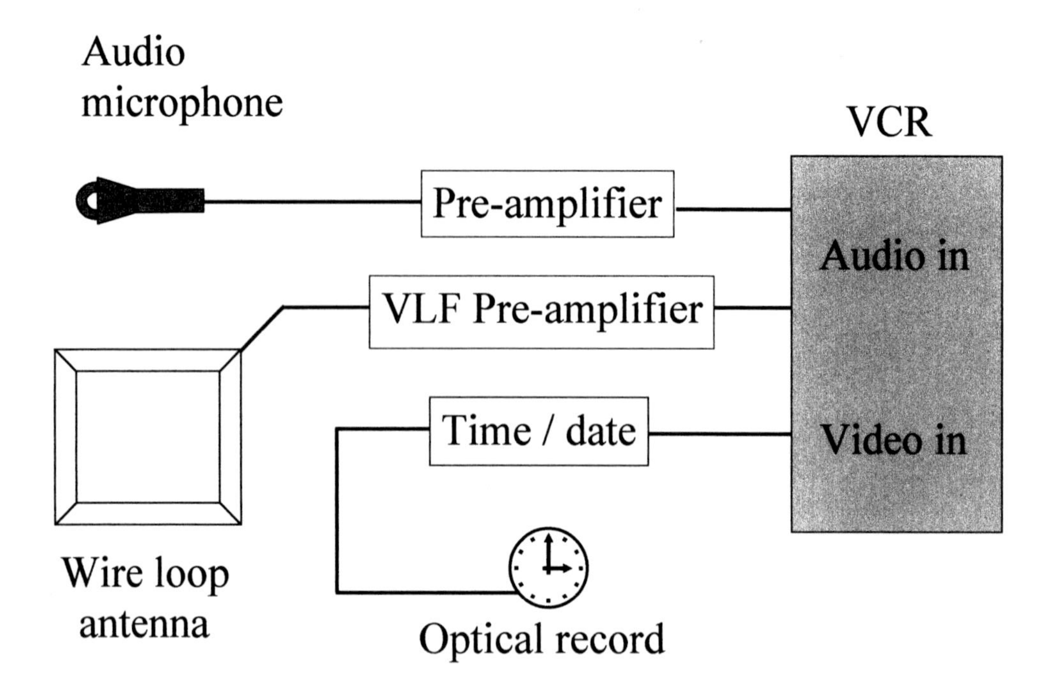

Fig. 59 A schematic diagram of a sound survey system. A circuit diagram for the VLF pre-amplifier is shown in Figure 60. The pre-amplifier for the audio microphone can be purchased from any audio or general electronics store.

precede the fall of a meteorite. There are both historical and contemporary accounts, for example, of bright Perseid, Lyrid and Leonid fireballs (all derived from soft cometary material) producing both sonic booms and simultaneous sounds. In a recent survey I made of fireball reports for 1962 to 1985 gathered from across Canada, I found that about one in every thirteen – that is, about 7.5 per cent – of bright fireball events are accompanied by sound phenomena (for details of the fireball archive and an analysis of the data, see http://hyperion.cc.uregina.ca/%7Eastro/MIAC/MFA/Intro.html). Further, the survey indicates that when a fireball is accompanied by a sonic boom, 13 per cent of the eyewitnesses actually hear the boom at a level sufficient to comment upon it. But in instances where simultaneous sounds were reported to accompany a fireball, only 6 per cent of eyewitnesses who submitted reports mentioned having heard them.

It is known that only very bright fireballs produce sounds, and a minimum peak brightness of about magnitude –6 to –8 is required before sounds of any kind are likely to be heard. The reason for this is reasonably straightforward: to generate sound, a meteoroid has to penetrate to an altitude of 30 to 40km, to achieve this a meteoroid must be large to begin with, and big meteoroids produce bright fireballs.

A Fireball Sound Survey

In the realm of rare recordings, those of sonic booms and simultaneous sounds must rank amongst the most precious. Only a handful of events have ever been recorded, and all of them by accident rather than design. The very first recording of a fireball's sonic boom was made serendipitously by an ornithologist on 25 April 1969. The boom was associated with the fall of the Bovedy meteorite in Northern Ireland. While the chances of hearing fireball-related sounds are very small, they are not zero, and one long-term survey project that the keen meteor observer could contemplate is to attempt to record such sounds. While such a project will never produce large quantities of fundamental data, just recording one event would be a significant triumph – the capture of one of the rarest sounds in nature.

Very little technical equipment is required for a fireball sound survey. Simplicity and reliability are the two main points; the equipment must be functional at all times (if possible) and should require the absolute minimum of maintenance. If one has to continually upgrade, coerce and tinker with the survey equipment, then the project soon becomes a proverbial 'pain'. If the equipment is simple and robust, there is no reason why it cannot simply be left alone to do its job with the barest minimum of attention.

A 'bare bones' fireball sound survey system can be made from of a simple wire loop antenna, to detect the VLF component of any potential simultaneous sounds, and a standard audio microphone. The output from the antenna and the microphone can be recorded on the right and left audio channels of a standard VHS videotape, while the time and optical signal can be recorded on the video channel. Figure 59 shows a schematic view of the survey equipment (the video component of the survey is described in Chapter 6).

The wire loop antenna for detecting the magnetic field component of the VLF radiation is made of thin antenna wire wound around a square former. The area of the antenna frame should be about a square metre, so the first step is to construct a square wooden or plastic frame with sides of about a metre. Antennas larger than this are technically more sensitive, but there comes a point where the whole antenna is too unwieldy – keep the frame small and suited to the space in which it will eventually be positioned (*see* below). Not only this, but the more sensitive the antenna, the more likely it is that the system will suffer from interference and the background hum caused

Fig. 60 *A circuit diagram for a straightforward VLF preamplified system. Components: R1 = 390kΩ, R2 = 10kΩ, R3 = 330kΩ, R4 = 6.8kΩ; C1 = 10μF electrolytic, C2 = 5μF electrolytic, C3 = 0.02μF disk, CA = 0.02μF disk (to be placed across the antenna terminals); T1 = T2 = general-purpose germanium transistor, e.g. 2N4314; 6V power supply, audio output jack. (Adapted from a design by G. Drobnock,* Sky & Telescope, *March 1992, p. 329)*

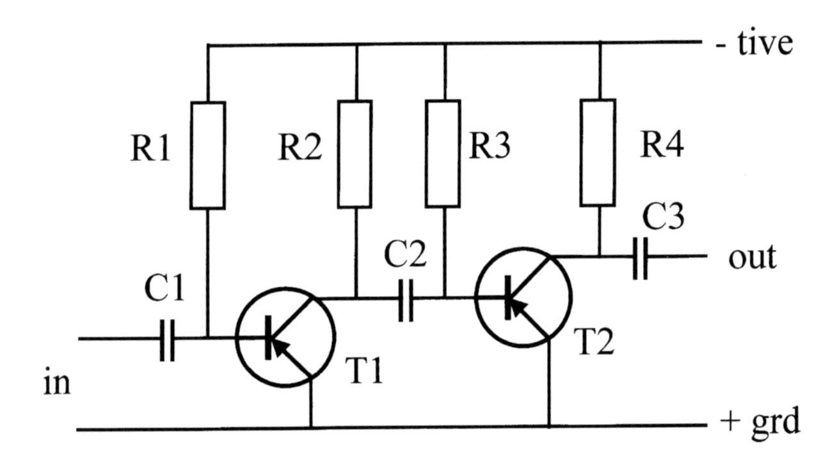

by electrical power lines. Because electrophonic sounds can be heard by human beings, systems with only a little amplification and sensitivity should be adequate for detecting any VLF signal. Once the frame has been constructed, winding a length of insulated 31-gauge antenna wire around the frame's exterior completes the antenna. The number of turns of wire is not that important, although the more turns, the more sensitive the antenna. There should be at least a few tens of turns, but more than two hundred turns is probably unnecessary.

A diagram for a preamplifier circuit suitable for a VLF monitoring system is shown in Figure 60 (*previous page*). All the components are readily obtainable from any electronics supplier, and can easily be assembled on a standard electronic prototyping board. To help tune the circuit, a 0.02μF disk capacitor can be soldered across the antenna terminals before they are attached to the pre-amplifier. The circuit provides a broadly tuned VLF receiver in the frequency range from about 5 to 15kHz.

Once constructed, the antenna should be mounted and secured in a vertical position. It need not be located outdoors – the loft will do nicely – but an outdoor antenna is probably less in the way and probably easier to access. And if the antenna is located indoors, the VLF system will probably suffer from considerable interference from domestic appliances – experiment with the antenna position, and see what works best. The antenna can face in any direction as long as its plane is vertical, since fireballs can occur anywhere in the sky. The antenna will be most sensitive in the plane of the loop, however, so to produce a strong signal a fireball will have either to pass through the plane of the antenna or pass nearly overhead of the antenna.

The output from the audio microphone, for recording sonic booms, will also require some preamplification before being connected to the audio channel of the VCR. Such microphone preamplifiers are inexpensive and may be purchased ready-made from any general electronics or audio store. The audio microphone need not be located at the same spot as the wire loop antenna, and again, it need not be outdoors. Work with the space that you have available, but remember that if the audio microphone is to be located outdoors, then it will need some protection from the elements – wind, rain, snow and cold will all conspire to significantly reduce the lifetime of a microphone. Weatherproofing a microphone for outdoor use is relatively straightforward: make sure that it is waterproof, and if possible placed in a porch or under a roof arch for further protection.

Once the microphone system and loop antenna are completed and installed, the survey can begin. (Technically, the optical device should also be functional at this stage, but more on this later.) The output from the antenna and microphone are fed into the left and right audio channels of the VCR. Since ease of use is paramount in such surveys, I would recommend that for recording the data you use the longest-running VHS videotapes you can obtain (4 hours or more).

Keep a detailed logbook of the day, date and times when the system is operational. Make a note of any changes or upgrades to the equipment. Record the dates on which new tapes start being used. The videotapes should be replaced fairly often since their playback quality is degraded slightly with each use. Once the system is in routine operation, make sure that you keep track of local fireball events (i.e. via the internet and newspapers). You certainly will not want to review the tapes each day; instead, wait for an event to be reported and then listen through the appropriate tape for the reported date and time. You will have to be a little creative in checking a tape for sounds, but if you know the start time of the tape and the time of the event, then the approximate position on the tape to be checked can be

Fig. 61 A 2kg fragment of the Innisfree meteorite as it was found in the snow. The sunglasses give an approximate scale. Because the ground was frozen at the time of the fall, the meteorite did not penetrate the soil. (Image courtesy MIAC and the Geological Survey of Canada: http://miac.uqac.ca/MIAC/morp.htm)

estimated and fast-forwarded to. In Chapter 6, I discuss how a video camera can be used to add a time clock to the video tape. It can take several days for eyewitnesses to report their observations, so it is best to have more than three or four days' worth of videotapes. My experience is that at least a week to ten days' supply of tapes is required to 'catch' the majority of locally reported fireball events.

We live in a world where any number of odd sounds are generated all the time. One way of eliminating false signals is to build several recording systems and locate them at sites separated by as many kilometres as can be achieved between a group of observers. With well separated locations for the recording stations, one can search for common sounds and attempt to eliminate sounds heard at just one station. In addition, the delay between the arrival times of the sound signal at different locations can be used to determine the fireball's direction of travel. This calculation is, however, non-trivial in that the variation in sound speed and local weather conditions must be taken into account.

Meteorite-Producing Fireballs

The Meteorite Observation and Recovery Project (MORP) operated a network of twelve camera systems in the Prairie Provinces of Canada from 1971 to 1985. In these fifteen years over a thousand fireballs were photographed (see Figure 98) by at least two of the camera stations, and 44 of these fireballs events produced meteorites with masses between 0.1 and 11kg. The most striking MORP success was the recovery of the Innisfree (Alberta, Canada) meteorite, which fell on 5 February 1977.

Canadian astronomers Ian Halliday, Alan Blackwell and Arthur Griffin, of the Herzberg Institute of Astrophysics, analysed the characteristics of the fireballs photographed with the MORP camera and derived the properties of a typical meteorite-producing fireball. The deduced characteristics are given in Table 4. The typical duration of a meteorite-producing fireball is between 3 and 5 seconds. The duration, as might well be expected, is strongly correlated with the zenith angle of the fireball, with longer-duration fireballs corresponding to larger zenith angles. The typical beginning and end heights of the luminous trail are 72 and 31km respectively, and while the beginning height does increase with increasing initial speed, the data indicate that there is no strong correlation between the end height and the initial speed. Interestingly, the height of maximum brightness doesn't appear to correlate with the initial speed of the meteoroid. The typical initial speed for meteorite-producing bodies is about 15km/s, but speeds as high as 27.9km/s have been recorded. The typical maximum brightness of the meteorite-producing fireballs was magnitude –9.2, although a range of –7 to –12 was recorded. The upper brightness limit (magnitude –12) is comparable to the brightness of the full Moon. Brightness alone is not a good indicator of whether a meteorite is likely to fall, but the observations do reveal that a terminal speed (the speed at the end of the luminous trail) below 10km/s is a very good indicator that a meteorite fall has taken place.

Many of the fireball images studied by Halliday, Griffin and Blackwell showed the presence of trailing fragments, which revealed that the parent meteoroid had broken apart. By looking specifically at the fragmentation height data, they concluded that meteoroids capable of producing meteorites probably contained pre-existing cracks, or are not solid all the way through. Indeed, the MORP study suggests that while it is ablation and initial speed that limit the potential survival of low-mass meteoroids, it is the pre-existing cracks within a meteoroid that determine the maximum mass that can survive to reach the ground.

Table 4

Median (most common) values of observed meteorite producing fireball characteristics. (Data from Halliday, Blackwell and Griffin, *Meteoritics*, vol. 24, pp. 65–72, 1989)

Quantity	Median value
Duration	4.2 seconds
Zenith angle	51.0°
Beginning height	72.0km
End height	31.0km
Initial speed	15.2km/s
Terminal speed	8.2km/s
Maximum brightness	mag. –9.2
Height of maximum brightness	47.5km

4 Collecting Meteorites

Meteorites fall to the ground all the time, day and night, without let-up. Something like 4,500 meteorites more massive than 1kg land on the Earth's surface each year. The majority of these meteorites, of course, will never be found, mainly because oceans cover most of the planet. And even if a meteorite does fall on dry land, it is often only good fortune that determines whether it will actually be found. Falls in remote locations far away from cities and towns, or in mountainous and wooded regions, are unlikely to be recovered, even if the atmospheric path of the associated fireball is known.

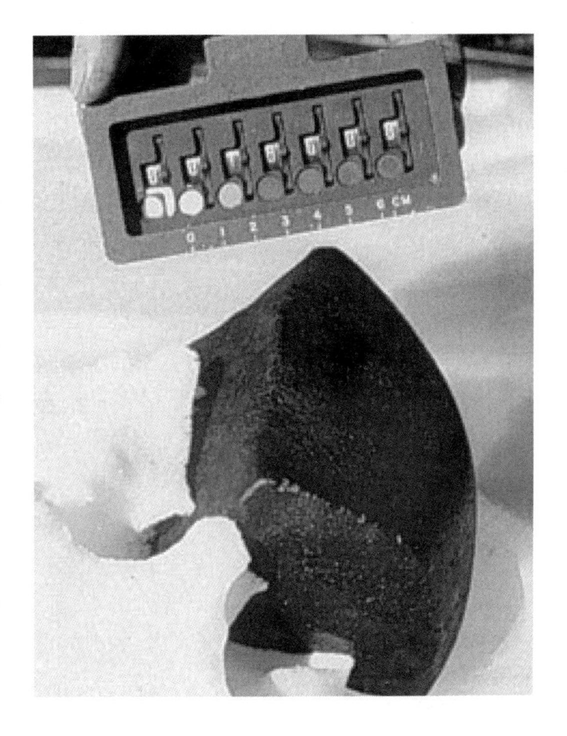

Fig. 62 Exceptions to the rule – finding meteorites in Antarctica. The cold desert environment in Antarctica is perfect for preserving meteorites, and the accumulation of meteorites by glacial flows has resulted in the formation of several meteorite-rich ice fields. Here a field researcher has found a small meteorite in the process of being exposed at the ice surface. (Image courtesy ANSMET)

Falls and Finds

Meteorites are generally described as being either falls or finds. If a meteorite is a **find**, this means that the exact date of its arrival on Earth is unknown. If a meteorite is classified as a **fall**, though, then its exact arrival date is known, which usually means that its associated fireball was seen. As of December 1999, the number of meteorites that had been collected and documented amounted to some 40,315 specimens*, of which 1,005 were falls. Of the others (the finds), 17,808 had been collected from Antarctic ice fields (*see* Figure 62). A meteorite fall is potentially more important than a find because the known arrival time, combined with observations of its associated fireball, can allow its orbit to be determined.

** These statistics are taken from the Catalogue of Meteorites (CUP) prepared by Dr Monica Grady at the Natural History Museum in London.*

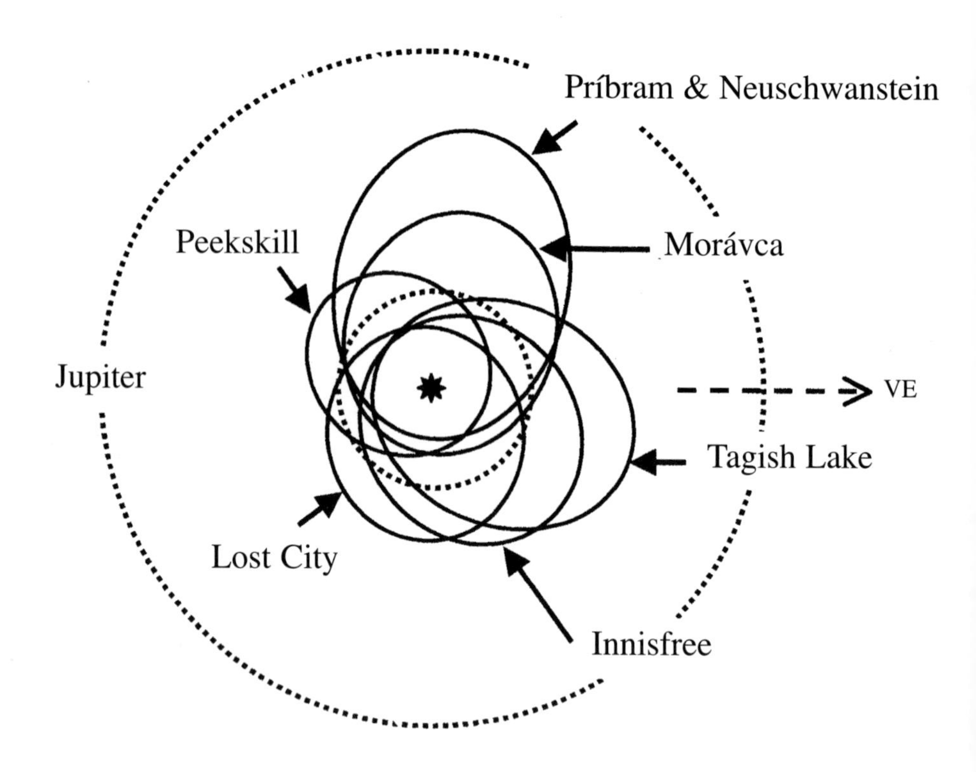

Fig. 63 *Orbits derived for seven meteorite falls, projected onto the plane of the ecliptic. They all have aphelia in the main asteroid belt, between the orbits of Mars (inner dashed circle) and Jupiter (outer dashed circle). VE is the vernal equinox.*

When we know the orbit of a meteorite, we know where it came from within the solar system, and its origins can be investigated. The orbits of seven well-studied meteorite falls are shown in Figure 63, from which it can be seen that they all have aphelia in the main asteroid belt, between Mars and Jupiter.

While an association between meteorites and asteroids can be established on the basis of their observed orbits, one particular group of meteorites, the so-called HED achondrites (from the names of three classes of meteorite, howardites, eucrites and diogenites), can be directly linked to a parent asteroid. On the basis of reflectance spectroscopy studies, it has been determined that all HED achondrites originated on the asteroid Vesta.

While the vast majority of meteorites that have been collected are derived from the main-belt asteroid region, some 84 meteorites are now recognized as being from the Moon, and a further 37 are known to be from the planet Mars. These meteorites represent material that was literally blasted into space as a result of asteroid impacts on the surfaces of the Moon and Mars. The lunar meteorites are partially recognized through their distinct composition, identifiable with the lunar rocks brought back to Earth by the Apollo Moon missions. The origin of the Martian, or so-called SNC (pronounced 'snick') meteorites, was established by studying the gases trapped in their interiors, which is an exact match with the composition of the Martian atmosphere, first measured by the Viking landers in the mid-

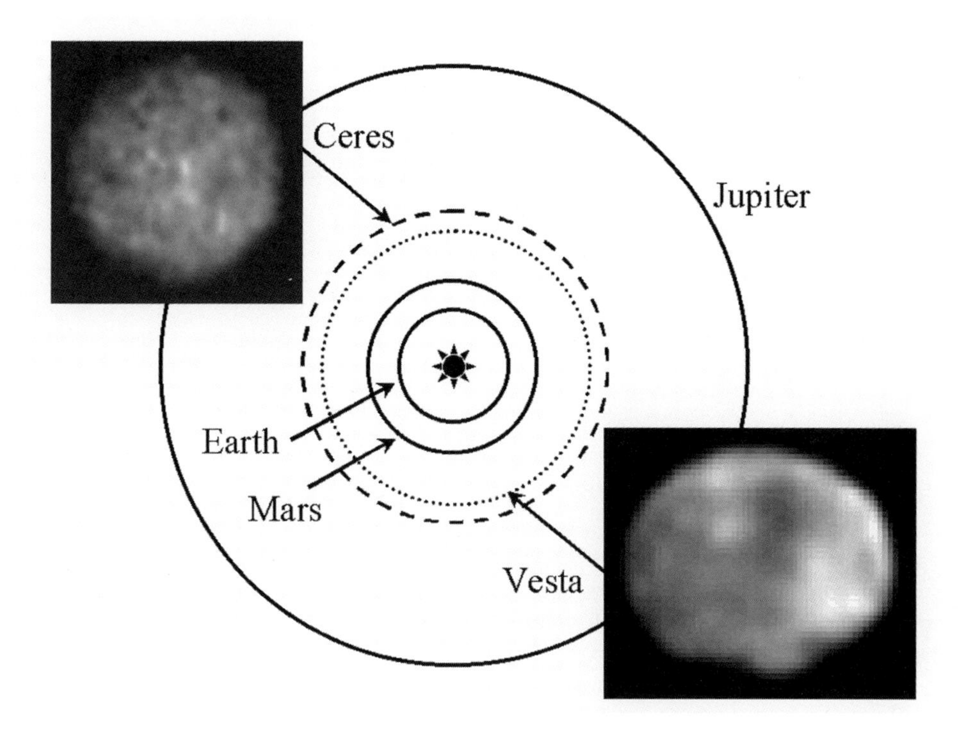

Fig. 64 *A schematic diagram of the orbits of the asteroids Vesta and Ceres. Vesta, as imaged by the Hubble Space Telescope, is about 500km across and has a highly reflective surface. Indeed, it is the only asteroid that can, under good viewing conditions, be seen with the naked eye. Ceres, the largest asteroid and the first to be discovered, is some 950km across and is possibly one of the parent objects of the so-called carbonaceous chondrite meteorites. (Images of Vesta and Ceres courtesy NASA/STScI)*

1970s. (The SNC prefix is derived from the first three falls recognized as being from Mars: Shergotty (India, 1865), Nakhla (Egypt, 1911) and Chassigny (France, 1815).)

Meteorite Classification and Origins

Meteorites can be physically divided into two distinct groups: the undifferentiated (see below) or chondrite group of meteorites, which contains the stone (or ordinary) and carbonaceous subtypes, and the differentiated group, which contains the irons, stony irons and achondrite subtypes. Since meteorites are derived from asteroids, the different meteorite types that are recognized must have come about because asteroids themselves have undergone internal processing, or differentiation, during the course of which low-density materials (e.g. silicates) have separated out from high-density ones (e.g. nickel–iron). A basic meteorite classification is shown in Figure 65. The undifferentiated group of meteorites shows the least amount of heat processing (i.e. melting), and the relative abundance of elements (excluding hydrogen and helium) in these meteorites is the same as in the Sun. Most interestingly, the carbonaceous chondrites contain organic material and minerals that can have formed only through chemical reactions involving water. Meteorites in the differentiated group have been greatly altered by heating. The stony achondrites

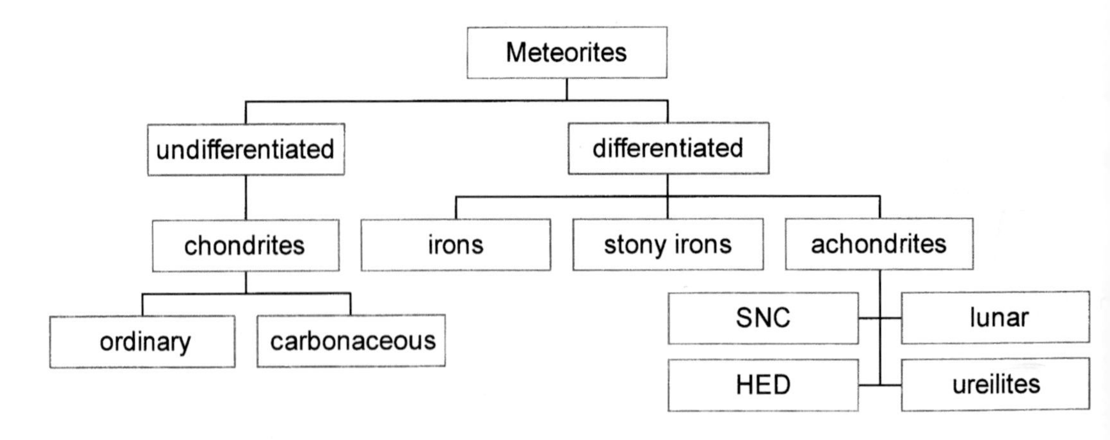

Fig. 65 *Schematic classification of the meteorite types that are discussed in the text. The ureilite achondrites are perhaps the least well understood group of meteorites, and intriguingly they contain carbon-rich material in the form of graphite and diamonds.*

resemble terrestrial basalt rocks, whereas the irons have a predominantly nickel–iron composition. The stony irons are meteorites that have a near half-and-half mixture of nickel–iron and silicate material.

In terms of their origins, it is believed that the undifferentiated meteorites are derived from asteroids that were never hot enough to undergo differentiation. If, on the other hand, an asteroid has undergone sufficient internal heating (e.g. through the decay of radioactive elements), then its interior might have partially melted. In this manner a core-and-mantle structure can develop. The core contains the higher-density elements such as nickel and iron, while the mantle contains mainly lower-density silicate material (see Figure 66). It is believed that many of the iron group of meteorites are samples of core material from asteroids subsequently fragmented by collision, and that the achondrites are samples of mantle material. The stony iron or pallasite meteorites are thought to be fragments from the core/mantle interface region. Some of the iron group of meteorites and the so-called mesosiderite stony iron meteorites are thought to have formed as a result of collision-induced heating, from further impacts on the surface of the parent body.

Figure 67 shows that about 86 per cent of meteorite falls are ordinary chondrites (of which about 3 per cent are carbonaceous chondrites), 8 per cent are stony achondrites (of which about 70 per cent are from the HED group), 5 per cent are iron meteorites and 1 per cent are stony iron meteorites.

How to Identify Meteorites

By far the best way to learn how to identify meteorites is to handle them – indeed, inspect as many of the real objects as you can lay your hands on, and don't rely on just looking at photographs. Many national and local museums, as well as universities, have meteorite collections, and it is well worth approaching the curators of such collections to arrange a viewing. Once you have seen and handled a few meteorite specimens, their key features will begin to become clear.

Most obviously, a meteorite that has been recovered shortly after falling will have a distinct **fusion crust**. For stone achondrites the fusion crust is often the only initial clue that an object is indeed a meteorite. The fusion crust of a newly landed meteorite is typically a

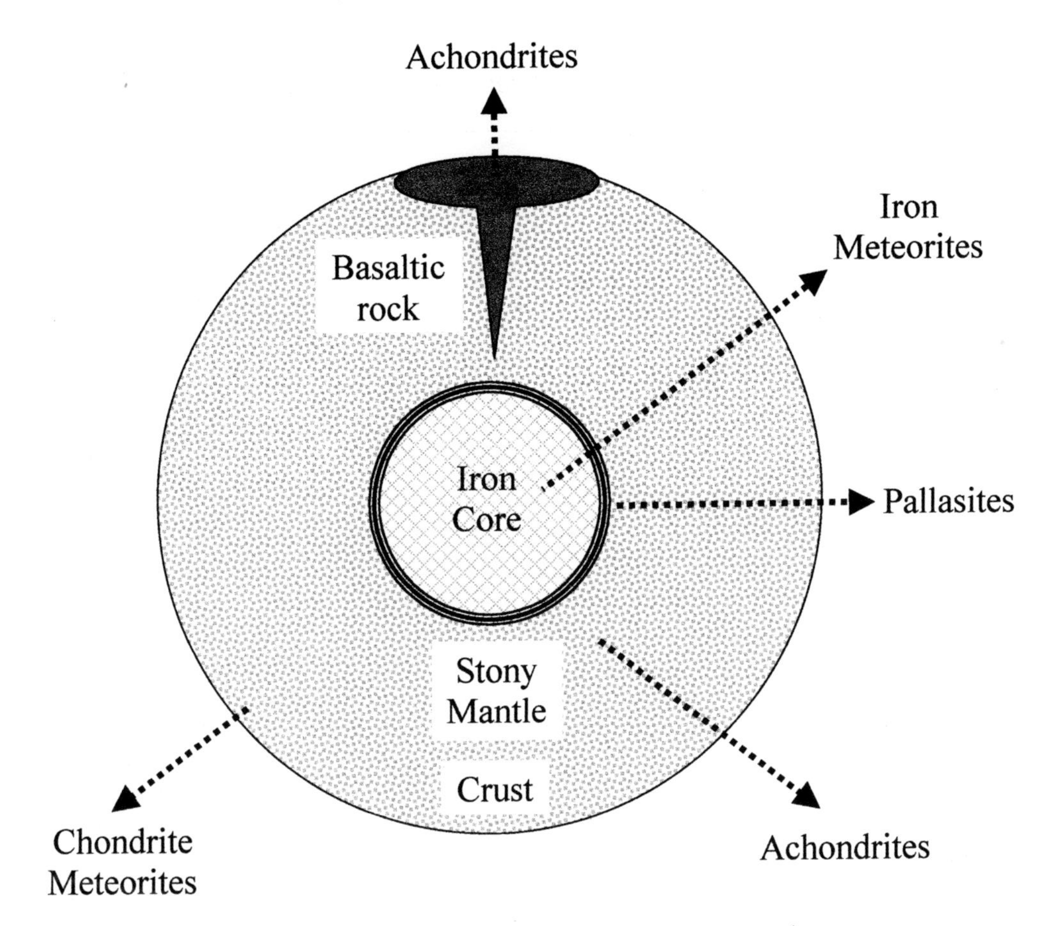

Fig. 66 Potential source regions of meteorites (schematic).

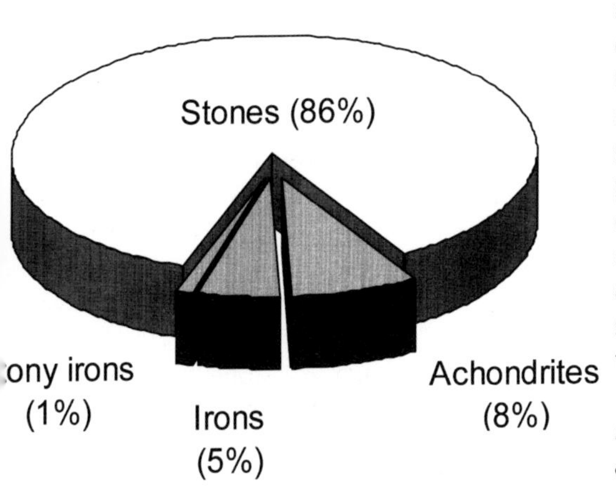

Fig. 67 The distribution of meteorite types according to fall statistics.

distinctive shiny black colour, and in stony meteorites this crust will stand out in sharp contrast to the light-grey interior (see Figure 68). The fusion crust, typically less than a millimetre thick, is composed of an iron-rich glass, solidified from the molten surface produced as the meteorite is heated during its descent through the Earth's atmosphere. Iron meteorites also have a fusion crust, but it is not always easy to distinguish. The fusion crust of stone meteorites will often show a distinctive trellis-work of thin, hairline cracks caused by

***Fig. 68** Flow lines of once molten material are clearly visible across the dark fusion crust of the Kirbyville (achondrite) meteorite that fell on 12 November 1906 in Jasper County, Texas. Hairline cracks can be seen in the fusion crust, and the lighter-coloured stone interior is visible on the far right of the image (see also the Morávka meteorite images in Figure 76). The longest dimension of the meteorite is about 50mm. (Photograph by Geoffrey Notkin; image courtesy Monnig Meteorite Gallery, http://monnigmuseum.tcu.edu/index.shtml)*

the rapid cooling that occurs during a meteorite's dark flight, immediately before it hits the ground. Both stone and iron meteorites can show what are called **regmaglypts** – indentations resembling thumbprints in the surface. Figure 69 shows the distinctive regmaglypts on the Mazapil iron meteorite (*see also* Figure 73).

A stony meteorite is difficult to identify if its fusion crust has weathered away, but some key features will remain to distinguish it from 'ordinary' terrestrial rock. First, the chondrites can be identified from their interiors, which contain many small spheres ranging in size

Fig. 69 *Regmaglypts on the surface of the Mazapil iron meteorite, which fell in Mexico on 27 November 1885. (Image courtesy the Naturhistorisches Museum, Vienna)*

from a few tenths of a millimetre to perhaps as big as 2mm across. Indeed, these spheres – called chondrules, after the Greek word for 'seed' – are the defining characteristic of the chondrite meteorites (*see* Figure 70). A small hand-lens will indicate whether a candidate rock has any distinctive spherical chondrules embedded in it. Some terrestrial rocks (e.g. oolitic limestone) also contain spherical structures in their interior, so in order to be able to identify a meteorite on this basis it is essential to be familiar with the characteristics of as many types of rock as possible. In addition to containing chondrules, stony meteorites also contain small nickel–iron grains. If one files a small section of a suspect rock's surface and small metallic flecks are produced, the rock may be a meteorite. The presence of nickel–iron grains within ordinary stone chondrites also makes them magnetic. A very good first test of any suspect rock, therefore, is to see if it will attract a magnet or deflect a compass needle. If a suspect rock is not

Fig. 70 *A cut section of a carbonaceous chondrite meteorite shows the distinctive chondrule structure of its interior. The slice is some 44 × 42 × 5mm in size. The meteorite is one of the thousands of fragments that fell near Allende, Mexico, on 8 February 1969. (Image courtesy the Hartmann Meteorite Collection, http://www.meteorite1.com)*

magnetic, or it attracts a magnet in just a few spots, it is probably not a meteorite.

Iron meteorites tend to stand out because of their odd shape and distinctive weight. One feature that is unique to iron meteorites, however, and one which is never seen in industrial iron, is the intersecting network of lines known as the **Widmanstätten pattern** (Figure 71). These lines are visible only after a cut surface of a sample has been polished and etched.

All iron (and stony chondrite) meteorites contain some nickel, and this can be chemically tested for. First, a small sample (just a few flecks will do) of the suspected meteorite

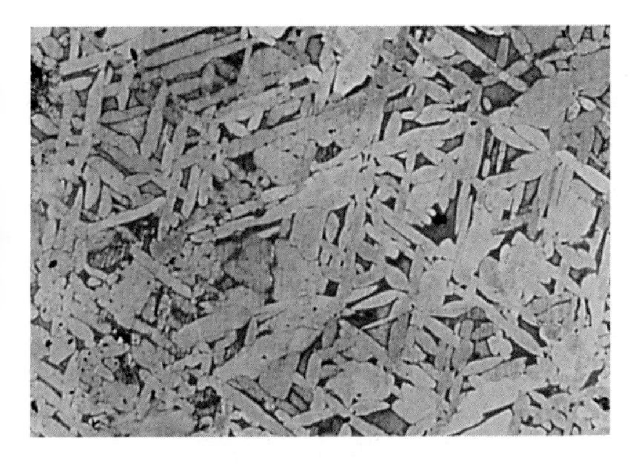

***Fig. 71** The Widmanstätten pattern of crisscrossing lines is clearly visible in this photograph of a polished and etched slice of the Willow Creek iron meteorite. The meteorite was found in Wyoming, USA, in 1914.*

should be dissolved in a solution of dilute hydrochloric acid. Once the sample has dissolved, the acid mixture should be neutralized with ammonium hydroxide. Filter out any precipitates and then add a few drops of dimethyl glyoxime to the mixture. If nickel is present in the solution, it will turn a bright cherry red.

Searching for Meteorites

Although meteoritic material is scattered across the Earth's surface, the chances of actually discovering a sizeable meteorite are staggeringly small. Very few people have ever been fortunate enough to come across a meteorite find by accident or to witness a meteorite fall. If you do fancy having a go at searching for meteorites, the key point to remember is that they can fall anywhere at any time, so it is always worth looking no matter where you happen to be. A planned meteorite search should, however, be organized carefully in advance. Since searches will often be conducted outdoors and perhaps in remote locations, safety issues should be of paramount concern. Check local weather forecasts, be prepared for all types of weather, carry and use a GPS receiver to monitor your progress, arrange to make progress calls at specific times and stick to a pre-arranged schedule. Once all such issues have been addressed, then the search can begin. While any location may yield meteorites, regions that are heavily wooded, marshy or covered with surface stones are probably not worth searching – any meteorites in such areas will be very difficult to locate. Ideally, you should search in wide open spaces that are easy to traverse.

A meteorite search should be conducted in much the same way as a search-and-rescue mission. Cover the area of interest thoroughly and methodically. Use local landmarks for your guidelines and simply walk back and forth scanning the ground in front of you. Repeat walking up and down until the entire search area has been covered. Hopefully, at some stage a meteorite fragment will be found. Metal detectors have been used to find meteorites – even stony meteorites will produce a 'target' signal, but how useful detectors are will depend very much upon the surface terrain and upon the prevalence of metallic fragments left over from past industrial activity.

If you do find a meteorite, well done! Make sure that you record a few important details. It is very important that the find location is well documented. Use a GPS receiver or a large-scale topographic map to determine as accurately as possible the latitude and longitude of the site. Make a note of the discovery conditions, the soil type, the vegetation, the rock cover and any other features of the site that may be useful in its future identification. Record the mass of the meteorite and take photographs of it in situ. Handle any freshly fallen meteorite as little as possible; wrap it in tin foil, and then place it in a plastic bag for protection. Once you have removed the meteorite from the site you will have to contact a recognized laboratory or

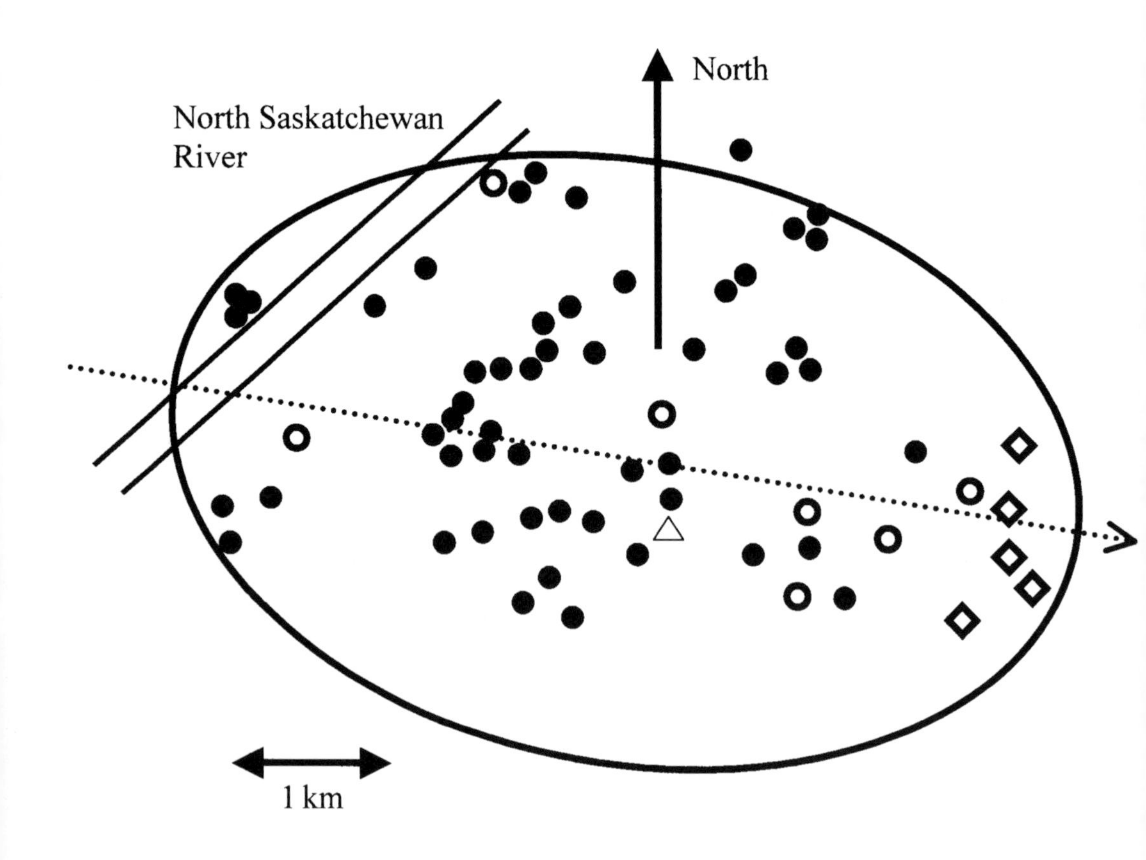

Fig. 72 *The Bruderheim meteorite strewnfield. The fall ellipse indicates a reasonably steep angle of atmospheric entry, and the direction of travel was from left to right in the diagram. The solid circles represent the locations where fragments less than 4kg in mass were found. The open circles show where meteorites with masses between 4 and 24kg were found. The open diamonds indicate where the largest meteorites, with masses greater than 24kg, were found. The larger, more massive fragments fell farthest down range. The open triangle near the centre indicates a triangulation point at latitude 53° 54″ N, longitude 102° 53″ W. (Diagram adapted from R. E. Folinsbee and L. A. Bayrock,* Journal of the Royal Astronomical Society of Canada, *vol. 55 (5), p. 218, 1961)*

museum to have it correctly classified – this is a non-trivial exercise and is best left to an expert mineralogist.

One reason for recording the precise latitude and longitude of any new meteorite find is that a large meteoroid will often fragment as it descends through the Earth's atmosphere. If you find one meteorite, then several more fragments may well be lying in close proximity, forming what is called a **strewnfield**. Depending upon the angle at which the meteorite entered the atmosphere, and at what height it broke apart, the strewnfield can be anything from nearly circular (for an almost vertical entry) to a highly elongated ellipse (for shallow angles of entry). The more meteorite fragments that are recovered, the better the strewnfield can be mapped out. In addition, the length of the fall ellipse's major axis and its

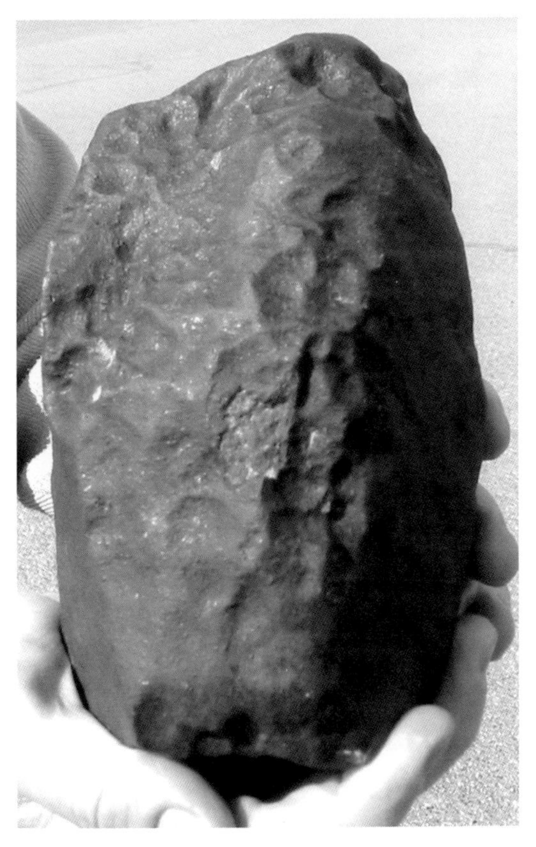

Fig. 73 *A large fragment from the Bruderheim meteorite strewnfield. Note the distinctive regmaglypts and the dark fusion crust. (From the University of Alberta's meteorite collection)*

orientation will give an indication of the atmospheric entrance angle of the meteorite, and the fact that larger meteorites will travel farther down range than smaller ones will give an indication of the direction of motion. Figure 72 shows the fall ellipse and strewnfield for the Bruderheim meteorite that fell in Alberta, Canada on 4 March 1960. Nearly 700 fragments were collected from this particular strewnfield, totalling 303kg in mass. The largest fragment weighed in at 31kg. The fall ellipse measures 5.6 by 3.6km.

When a meteorite falls at a shallow angle to the horizon, the fall ellipse can be very long. The Morávka fall in the Czech Republic on 6 May 2000 moved on a trajectory that was just 20° to the horizon, and consequently produced a fall ellipse that was nearly 30km long. Six ordinary chondrites with a total mass of 1.4kg were eventually discovered in the fall ellipse, and in a number of instances observers reported hearing a distinct whooshing sound, indicative of a fall nearby, but were unable to locate the actual meteorite fragments.

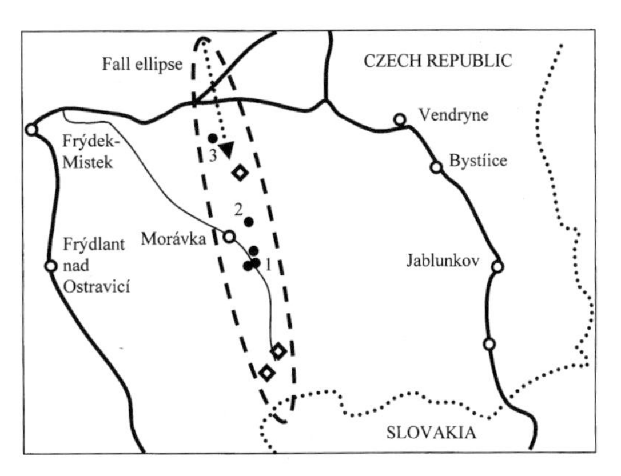

Fig. 74 *The fall ellipse for the Morávka meteorite. The solid circles indicate where meteorite fragments were recovered. The open diamonds indicate where sounds of a falling body were heard but no actual meteorite fragment was recovered. North is to the top of the diagram, and the arrow indicates the ground track of the fireball as recorded by video observations (see Figure 56). The numbers refer to the meteorite fragments shown in Figure 75. (Details of the Morávka fall can be found in the journal* Meteoritics and Planetary Science, *vol. 38 (7), p. 975, 2003)*

Collecting Micrometeorites

Micrometeorites, in spite of their smallness, are in fact more easily collected than centimetre-sized meteorite fragments. To start a collection of micrometeorites, all you need to do is leave a container outdoors for at least a few weeks.

Fig. 75 *Three Morávka meteorite fragments. The location at which each fragment was found is shown in Figure 74. Fragment 1, which has a mass of 214g, is to the left; fragment 2, which has a mass of 330g, is shown to the right. Fragment 3, at bottom left, has a mass of 91g. (Image courtesy J. Borovicka, Astronomical Institute and Academy of Sciences, Ondrejov, Czech Republic)*

The only other equipment required is a tea strainer (or a fine sieve), a piece of white paper, a strong magnet, a pair of tweezers, a fine-bristled paintbrush and a microscope.

Micrometeorites, by definition, are so small that they are not completely destroyed by ablation upon entering the Earth's upper atmosphere. Because they are so small, they are rapidly decelerated by the Earth's atmosphere and consequently take many days or even weeks to drift down to the ground. At the Earth's surface, micrometeorites accumulate on any exposed surface.

Material can be collected either by placing a large open-topped container in a location which is reasonably secluded, but open to the sky, for several weeks, or by gathering rainwater that runs off the roof of a house. Once you have collected a good sample of dust and debris, the first thing to do is pick out the larger, grit-like particles with a pair of tweezers. Next you will want to filter out the small dust particles – and this is where the tea strainer comes into play.

Add the dust and debris grains to a small quantity of water, and filter out any dirt or mud particles with the tea strainer. The very fine particles that pass through the strainer should be collected and allowed to dry out thoroughly. Once completely dry, carefully spread the residue, with a small paintbrush, on a piece of white paper. Now, by moving a strong magnet underneath the paper (try a neodymium magnet, since they are the most powerful magnets easily available), carefully separate out any magnetic material. Some, but not necessarily all, of the magnetic particles separated out will be micrometeorites.

Spread the magnetically separated particles on a clear glass microscope slide (with a fine-hair paintbrush), and examine the individual particles carefully. You will need to use a relatively high magnification, but not so high that you cannot make out the general shape and colour of the particles. The micrometeorites in your sample will typically be spherical in shape, showing perhaps a few surface pockmarks, and they will typically have a blackish coloration (Figure 76; *see also* Figure 6). Separate out any such particles and carefully save them; the remainder will be terrestrial grains or some form of industrial pollutant. If you have been lucky you may find a few micrometeorites at each 'harvesting' time.

Fig. 76 *Micrometeorite spherules collected in Antarctica near the Amundsen-Scott South Pole Station. The individual spheres are a few tenths of a millimetre across. (Image courtesy Dr S. Taylor, US Army CRREL)*

A general study of windborne material, including the collection of cosmic spherules, is being conducted at the University of Rhode Island, in the United States, as part of the National Aeolian Detritus Project (NADP). The project has an interesting website at http://geo.uri.edu/skydust/.

5 Visual Meteor Observations

The Sun has sunk below the horizon, and a clear, crisp and moonless evening is unfolding before you. With the promise of a perfect night for surveying the heavens, your thoughts turn to meteor counts, radiant locations and limiting magnitudes. Compared with many other branches of astronomy, meteor observing is a relatively straightforward procedure. Very little in the way of expensive equipment is required to begin observing, and scientifically useful data can be gathered after just a few nights' experience. All the data must be assessed and collated, however, and if anything useful is to come out of your observing session, systematic records must be maintained. There will be times when you will just want to sit back and enjoy the show, but if any observations of scientific value are to be made then a few simple rules need to be followed. This chapter is concerned mostly with the basic procedures that should be adopted before, during and after an observing session.

Join the Club

There are many local, regional, national and international astronomical societies around the world, many of which have a meteor observing group. Search out your local society and find out what projects the members are working on. Get involved. For the beginning observer, joining a local meteor observing group has many benefits since the established members will know good observing sites, and their experience will be of great benefit. In addition to a local group, you should also consider

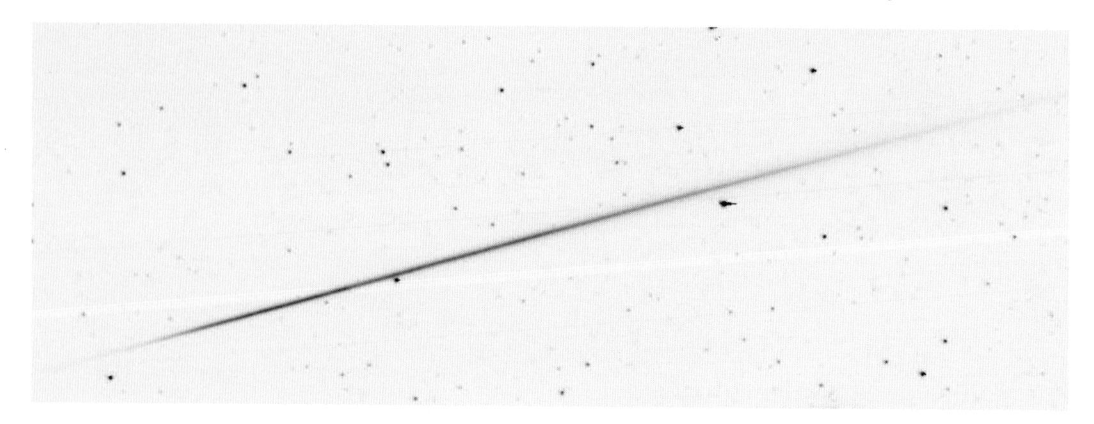

Fig. 77 *A bright Perseid meteor captured at the Rothney Observatory, University of Calgary, Alberta on the night of 12 August 2004. At the time I was seated in the observatory's control room, and distinctly recall hearing yells of recognition and delight from eyewitnesses outside the dome when the meteor sped across the sky.*

joining one or more of the national and international amateur astronomy societies. These organizations may be more remote in terms of where they meet, but they do provide valuable information on current and forthcoming events via the internet and society newsletters. Such organizations also tend to have extensive data archives and a host of people experienced in data analysis.

Read the Literature

By far the best way to keep informed about what other meteor observers are doing is to read and subscribe to the various magazines and journals dedicated to the topic. General news-stand magazines such as *Astronomy Now, Astronomy* and *Sky & Telescope* cover all topics related to astronomy, and are invariably a useful resource when it comes to forthcoming events. These magazines often publish finder charts for the radiant locations of the major annual meteor showers, and they periodically publish review articles on topics related to meteor observing. Your local library may well have back issues.

More technical articles on meteor observing and meteor physics are published in journals such as *WGN*, the *Journal of the International Meteor Organization* (IMO), and *Radiant*, published by the Dutch Meteor Society. The *Journal of the British Astronomical Association* (JBAA) and the *Journal of the Royal Astronomical Society of Canada* (JRASC) also publish occasional articles relating to meteor observing. Other organizations, such as the American Meteor Society, publish the observations gathered by their society members in regular newsletters. *The Astronomer* provides detailed information on meteor observing and other transient sky phenomena such as nova and comet sightings. The quarterly *Meteorite* is by far the best general magazine on meteorites and meteorite collecting. It publishes articles on meteorite searches, meteorite curation and the history of meteorite falls. In short, you should cast your reading net as widely and as deeply as possible.

Detailed research papers on meteor physics, the evolution of meteoroid streams and the analysis of meteorites are regularly published in professional astronomy journals. The keen reader may wish to consult, for example, the *Monthly Notices of the Royal Astronomical Society, Astronomy and Astrophysics, Planetary and Space Science*, the *Astrophysical Journal and Meteoritics & Planetary Science*. Back issues of most of the major astronomy journals can be accessed through the NASA-sponsored Astrophysics Data System (ADS) website at adswww.harvard.edu/.

Table 5

Recommended magazines and journals, and their related websites.

Astronomy Now	www.astronomynow.com/
Astronomy	www.astronomy.com/
Sky & Telescope	skyandtelescope.com/observing/objects/meteors/
WGN	www.imo.net/
Radiant	www.dms.org/
JBAA	www.britastro.org/info/meteor.html/
JRASC	www.rasc.ca/journal/
The Astronomer	www.theastronomer.org/meteors.html/
Meteorite	www.meteor.co.nz/

Preparations – Before You Go

As with any activity that will take you outside at night, away from populated centres and exposed to the vagaries of the weather, common sense and due diligence should be exercised at all times. As with most astronomical pursuits, meteor observing must be done at night and away from as much background light as possible. Indeed, the more remote and darker the site you can find to observe from, the better. But both these requirements put you at potential risk, so always plan ahead.

In daylight hours, check any location that you are planning to use as an observing site. Note the lay of the land – it is amazing how easy it is to fall into a ditch or stumble over a rock outcrop in the depths of a moonless night. If you do have an accident and injure yourself at a dark, remote location, then you should have some basic plan for survival, especially if you are observing on your own. An emergency first-aid kit and a warming space blanket are always worth carrying on any observing trip. Always carry a cellphone with you, but check beforehand that it works in the area you are going to observe from. Also, before you set out it is always worth telling someone what your observing plans are, and to indicate an intended time of return. It is easy to belittle such simple, commonsense safety concerns, but it is worth just a few extra minutes of preparation to ensure that any potential mishap can be dealt with.

Basic warm and waterproof clothing should be taken on all observing expeditions. Even in the summer you will soon feel the cold after a few hours of observing. In the winter you will be cold within minutes if you are not properly dressed. Dress warm, and have extra dry clothing available to change into. Also, even if it is the summer, take a pair of gloves and wear a hat. The hat will not only keep you warm, it will, in many climates, keep those 'itchy bugs' out of your hair.

Always take some food and drink on an observing expedition. The food will keep your energy reserves up, and the liquid will keep you hydrated and alert. Most observers, myself included, tend to ignore all slimming and health guidelines and consume large quantities of sugar-packed snacks and carbonated drinks when out observing. As the saying goes, 'It may not be pretty, but it works.'

Basic Equipment

You need no expensive equipment to begin visual meteor observing. There are a few essential items, however, that must be gathered together before the observing session starts. At a minimum you will require an accurate watch or clock that has an hours, minutes and seconds digital display, and a notebook and something to write with. The notebook should ideally have a hard cover and be able to withstand use in the field and being stuffed into a bag or pocket. If work of any lasting value is going to be done, then you must keep accurate and comprehensive notes. Each and every observing session must be recorded in detail, with notes being regularly jotted down on the observing conditions and the progress of the night's work. The notebook is your permanent record of where, when and what you observed. The more notes you make on each observing session, the easier it will be to reconstruct the conditions under which the data were gathered – especially if you need to review the data weeks, months or years after the actual observing session. While most people use a pen or felt-tip for note-taking, it is worth having some pencils as part of your basic kit. Ink has a habit of freezing in cold, wintry conditions – pencils work at all temperatures!

Many observers use a small voice recorder to make notes on the sky conditions and the characteristics of the meteors seen during an observing session. There is a wide choice of such recording devices available in every high-street electronics store, and in my own experience they are very useful. There are a few points to consider before purchasing a voice recorder, however. Remember that you may well be wearing gloves when observing, so buy a device with a voice actuated recording feature. Also, make sure that you have several spare sets of batteries and tapes when observing.

Comfort is essential during an observing session, so one of the most important pieces of equipment to buy before you begin observing is a camp bed or an air mattress. Some observers prefer to use a deckchair, but the point is that you will be spending many hours looking upwards at an angle of about 45° to the horizon, so your neck will need some support. It is generally advisable not to stand while observing since this soon leads to fatigue, affecting both the quality of the data collected and your enjoyment of the observing session. In my experience a camp bed with an angled headrest is perfect for observing, and allows observations to be made while you are in a sleeping bag. You are then insulated from the (generally cold) ground, your neck is supported at a good viewing angle, and the sleeping bag provides additional warmth.

In addition to the above items, you will need an ordinary torch/flashlight, as well as one with a red filter (to be used once you are dark adapted – *see below*) and a basic GPS receiver for determining the time as well as the latitude and longitude of your observing site. You will also need a good star atlas. Any atlas with large-scale charts showing stars down to at least magnitude +6.5 will serve. The International Meteor Organization provides an internet link to star charts from the *Atlas Brno 2000.0* which can be downloaded from http://www.imo.net/visual/minor/gnomic/atlas. Cambridge University Press also publishes a version of *Sky Atlas 2000.0* by Wil Tirion and Roger Sinnott with clear plastic laminated pages, which provides useful protection for the charts on damp nights. Additional items that you might take on an observing expedition are a pair of binoculars, and a laptop computer with associated astronomy and planetarium programs.

Beginning the Observing Session

Once you have arrived at your chosen observing site, it is important to allow some preparation time. Before you begin to gather data it is essential that you allow sufficient time to become dark adapted. While your pupil will begin to adjust very rapidly to the dark, it requires a good 20 to 25 minutes for the full photochemical adaptation of the eye to be achieved. It is essential that you avoid looking at any bright lights (e.g. a handheld torch/flashlight or even the headlights of distant cars) since just one short glance will destroy your dark adaptation. Use only a red light while making notes during the observing session. A red light is safe to use since the eye's dark adaptation is less sensitive to the red wavelengths of light.

Once your eyes are dark adapted, there are still a number of preliminary observations that must be made before you can settle down to count meteors. First, the general observing conditions should be recorded. Estimate what percentage of the sky is obscured by cloud, and also note how much of the horizon is obscured by trees or the local topography. These estimates are somewhat subjective, and you need not worry about distinguishing levels any finer than, say, 0, 5, 10, 25 and 50 per cent. The estimate of cloud cover is only intended to be a rough guide to how good the sky and observing conditions are. Clearly, if the cloud cover is more than 50 per cent then it is probably not worth beginning a serious meteor count. Keep a watch on the cloud cover as the night progresses and make a note of the percentage of cloud cover at regular intervals, or as often as seems necessary.

The Limiting Magnitude

In terms of evaluating the quality of the observing conditions, the basic measure of 'sky goodness' is the quantity known as limiting magnitude. The limiting magnitude (LM) is the magnitude of the faintest stars visible near the zenith. The LM will vary according to the transparency of the sky (which changes with elevation above the horizon), the presence of the Moon and background light pollution. It also depends upon the individual observer, in that the general level of perception will vary from one observer to the next. All in all, however, the darker and more transparent the sky, the larger the LM and the more meteors you should see. Under ideal observing conditions the normally sighted naked-eye observer can see stars as faint as magnitude +6.5. To detect fainter magnitudes requires either optical aid or photographic equipment.

There are a number of standard methods by which the limiting magnitude can be evaluated. By far the most straightforward is to count the number of stars that are visible

Fig. 78 *The IMO calibrated starfield in Cygnus for determining limiting magnitudes. This is starfield number 14, and the limiting magnitude is determined by counting how many stars can be seen within the designated triangle. In general one should make star counts in several calibrated starfields and determine an average limiting magnitude around the sky. Likewise, one should also use a calibrated starfield close to the direction that is being monitored for meteors.*

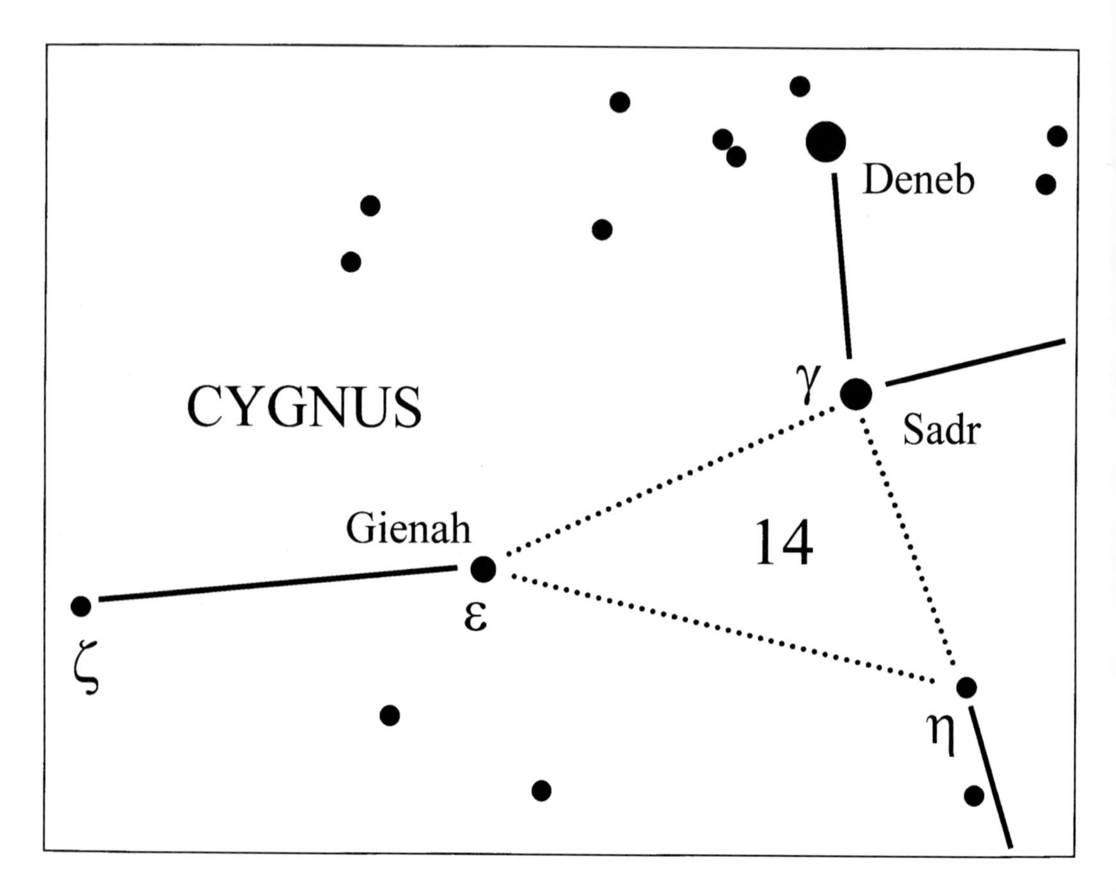

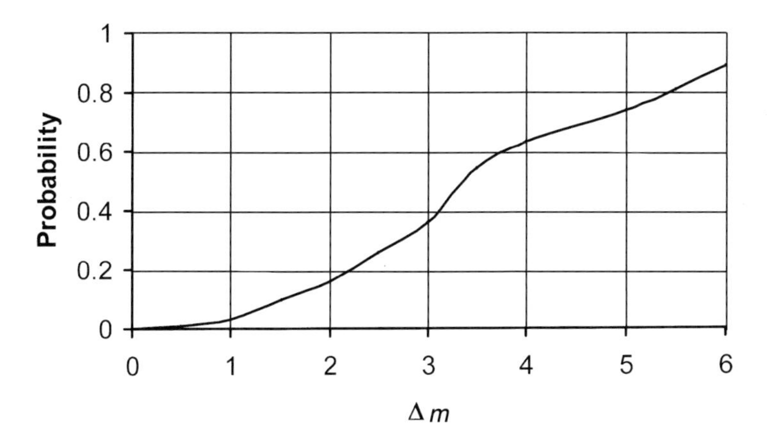

Fig. 79 *The probability of perception, P, plotted against the difference in magnitude between the limiting magnitude and the peak meteor magnitude, m: Δm = LM – m. (Diagram based upon data provided by Dr R. L. Hawkes, Mount Allison University, New Brunswick, Canada)*

within the boundary of a certain region of the sky. The International Meteor Organization (IMO), for example, has designated a number of standardized starfields for evaluating the limiting magnitude. An example is shown in Figure 78, which indicates the triangle formed by the stars Epsilon (e), Eta (h) and Gamma (g) Cygni. The LM is determined by counting the number of stars visible within this triangle. If you can see three stars within the triangle, then the LM is magnitude +4.0. If you can see seven, the LM is magnitude +5.0. Likewise, star counts of 14 and 24 correspond to LMs of +6.0 and +6.5, respectively. The entire set of calibrated IMO starfields can be viewed at www.imo.net/visual/major/observation/lm/. The North American Meteor Network (NAMN) has a web page that can be used to find calibrated LM starfields for particular observing sites: www.namnmeteors.org/lm_calc.html/.

One should evaluate the limiting magnitude at regular intervals throughout the observing session – and at least once every half-hour. The LM will change through the night, and in general if it gets brighter than +3.5, then there is little point in continuing to make any serious meteor counts.

Detailed studies of meteor count statistics obtained by experienced IMO observers have revealed, perhaps as one might expect, that the number of meteors recorded by any given individual is always less than the true number of meteors actually appearing. Even under ideal viewing circumstances, when the LM is magnitude +6.5, not every meteor appearing in an observer's field of view is going to be recorded. One can appreciate that this effect will be at its greatest for the faintest of meteors – those with peak magnitudes just above the LM. Studies have shown that the probability of perception, P, defined as the ratio $P = N/\phi$, where N is the number of meteors observed and ϕ is the actual number of meteors, varies according to the difference between the LM and the peak meteor magnitude, m. Figure 79 shows the probability of perception plotted against the magnitude difference, $\Delta m = LM - m$. From Figure 79 we can see that if a meteor is only one magnitude brighter than the LM, then there is only about a 3 per cent probability that it will actually be seen and recorded. The probability of perceiving a meteor exceeds 50 per cent only for meteors at least 3.5 magnitudes brighter than the LM.

What to Record During a Meteor Shower

If no particular shower is active on the night on which you are observing, then in principle you can look anywhere on the sky for meteors. If a shower is active, however, and you wish to monitor its activity, then it is recommended

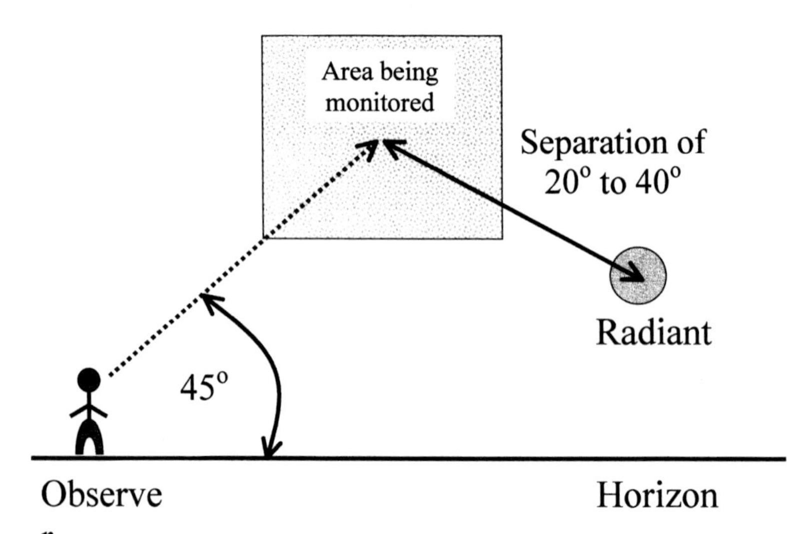

Fig. 80 *Where to look, with respect to the radiant, when observing a meteor shower. As a rough guide, the angle between the tips of your thumb and little finger when extended and viewed at arm's length is about 15°.*

that you concentrate on a region some 20° to 40° away from the radiant, where the meteor trails will be longer and more distinctive, and at an elevation of at least 45° above the horizon. If you plan to monitor the activity of a specific annual meteor shower, then you must first be sure that you know the location of the shower's radiant. Use a star chart to locate the radiant (see the Appendix) and fix in your mind's eye the general direction from which the shower meteors should be coming. Any meteor whose path cannot be traced back to the radiant should be deemed to be a sporadic meteor and recorded as such.

In monitoring the activity of a specific meteor shower, the times at which individual meteors appear is not especially important, and they need not be recorded. The priority is to record how many meteors are seen per unit time, typically in a one-hour interval. So focus your attention on the area of sky you have chosen to monitor, and try to minimize the amount of time you spend not looking at that region for meteors. In this respect the use of a voice-activated recording device is particularly useful – you can record your observations, limiting magnitude counts, and cloud cover estimates and any other details without taking your eyes away from the monitoring area.

When you see a meteor, there are number of details that you should assess both quickly and consistently. First, you will have to decide whether the meteor is a shower member or a sporadic. If it is a shower meteor, it should 'look' as though it emanated from the radiant (recall Figure 4, p. 11). Remember that the radiant is not a single point, but a region that is usually several degrees in diameter. If it is obvious that the meteor cannot reasonably be traced to a known active radiant, then log it as a sporadic meteor. In Figure 81, meteors A, B, C and D would be deemed to be shower members, whereas meteors E and F would be classified as sporadics. Remember also that the closer a shower meteor is to the radiant, the shorter its trail length will be.

After deciding on whether a meteor is a shower member or a sporadic, you need to estimate the meteor's maximum brightness. Being able to estimate magnitudes accurately (and consistently), to say half a magnitude, is a skill that takes some time to develop. Start by making sure that you are familiar with the magnitudes of the stars in the area of sky you are monitoring. Each time you go observing, find a series of stars of known magnitude and

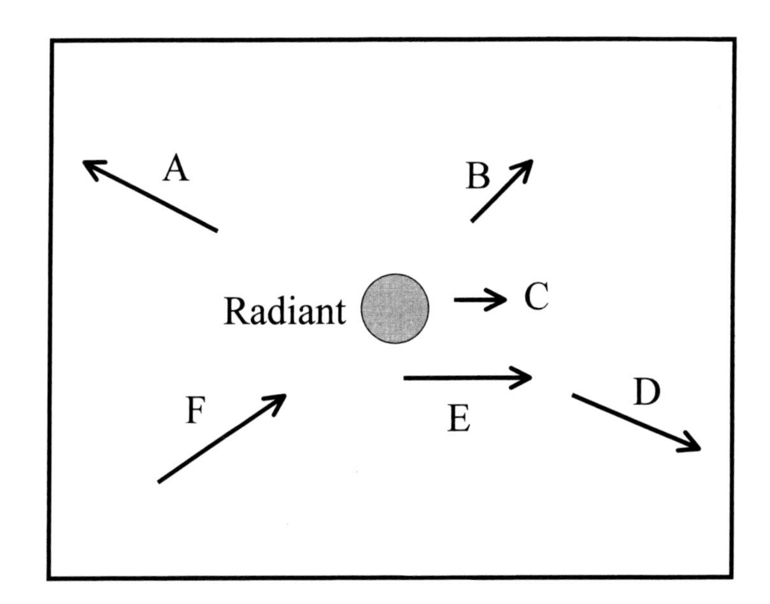

Fig. 81 *When a meteor shower is active its known radiant position can be used to gauge shower membership. Meteors A, B, C and D can all be traced back to the radiant region and would most probably be shower members. Meteor E is close to the radiant, but its long trail would indicate that it is probably not a shower member. Meteor E could also be excluded as a shower member on the grounds that its trail cannot be traced back to the radiant. Meteor F is clearly not a shower member because it is moving towards the radiant.*

try to form a mind's-eye picture of how they 'look'. Most of the conspicuous constellations are composed of stars with magnitudes between +4 and 0. If any planets are visible, they can be used as brighter comparisons, but remember that the brightness of a planet will vary according to its distance from the Earth. At their brightest, Venus, Mars, Jupiter and Saturn can reach magnitudes of –4.7, –2.8, –2.9 and –0.3, respectively. If you are fortunate to see a potentially meteorite-dropping fireball, then the Moon, if visible, could be a useful comparison. The full Moon is magnitude –12.7, and at quarter-phase it is magnitude –10.

Once a meteor's origin and maximum brightness have been established, you should record an estimate of its duration. This again requires some practice in order to get it right, but typically you are trying to estimate the time only to the nearest half-second. Most meteors are not likely to be visible for more than a couple of seconds; only a very bright fireball is likely to last for five seconds or more. With the duration estimated, the next feature to note is the meteor's general colour. Meteors' colours are somewhat subjective, and most will simply appear as a white streak, but some do have a distinctive red, blue or green colour. A faint, greenish persistent train is often seen in the path of fast meteors. If such a train is seen, an estimate of its duration should be made. Some meteors show short-duration enhancements, or flares, in brightness, and/or may fragment into two or more components. Other meteors may appear to flicker periodically in brightness. Such phenomena, if seen, should also be recorded. The reason for recording the colour and any flaring or fragmentation is that records of them enable a more complete picture of the meteor shower under study to be developed. In particular, they can help to reveal the characteristics of the meteoroids in the stream: flickering, for example, is though to indicate meteoroid rotation. It is always a good idea for a beginner to make observations in the company of an experienced observer and to compare estimates of duration, magnitude and colour and any other noted characteristics – this will help to hone the skills required for good reporting.

To illustrate how notes should be taken, here is part of the record made during a session of observing the Geminid meteor shower:

13/14 December 2004
Field is Leo Minor
LM 6.0, cloud 0
Begin: 22:15
Sporadic, 1.5, 0.5, green
Geminid, 0, 1.5, flickering
...
22:45. LM 5.7, cloud 5
Geminid, 3.5, 1
...
End, 23:30

The first note is of the night on which observations are being made, and next is a statement of what part of the sky was being monitored. The limiting magnitude is then given, as magnitude +6.0, along with a note of the cloud cover (0 per cent). The actual meteor count observations began immediately after making the first estimate of limiting magnitude, at 22:15 Universal Time. The first meteor seen corresponded to a green sporadic meteor of magnitude +1.5 with an estimated duration of half a second. It was followed by a zero-magnitude Geminid, lasting 1.5 seconds, for which distinct flickering was observed. At 22:45 UT further estimates were made of the limiting magnitude (slightly worse at +5.5) and the cloud cover (now 5 per cent). The next entry records the sighting of a magnitude +3.5 Geminid that lasted for 1 second. The last entry in the example is the time at which the observing session ended (23:30 UT). It is a good idea to record time markers every ten or fifteen minutes – even if the meteor count data are to be analysed over longer time intervals. The more data you have, the better you can 'reconstruct' the observing session at a later time.

Organizing the Data

Once the observing session has been concluded, then data analysis can begin. In most cases this will be on a later night than when the data were collected, which is why it is important to make clear and detailed notes during the observing session. The main aim at this stage of the analysis is to arrange the data into various time and magnitude 'bins'.

The numbers of shower and sporadic meteors seen during the succession of time intervals in the observing session are first extracted from the audio tape (or page notes). In this part of the analysis you are collating the number of shower as well as the number of sporadic meteors seen in each of the bins – the intervals between each pair of recorded time markers. You should also record the value of the limiting magnitude and the cloud cover for each bin. Usually the data need not be divided into time bins shorter than a quarter of an hour. The bin size to use will partly depend upon the circumstances of the shower. If only a very few meteors are being observed, then the counts from a successive string of time bins can be added together. In general, however, the total start to end time over which meteor counts might be added should not exceed one hour. If, on the other hand, unexpectedly large numbers of meteors are being seen during an observing session – as during a storm or outburst – or if the activity is varying rapidly, then one might choose time bins of 5 to 10 minutes. Such small bins can, of course, be

Table 6
An example summary table of observations collected during the Geminid meteor shower. In a total observing time of 2 hours and 35 minutes, 165 meteors were seen, of which 131 were Geminids and 34 were sporadic meteors.

Interval (UT)	LM	Cloud (%)	Geminids	Sporadics	Total
22:15–22:45	6.1	0	15	6	21
22:45–23:15	5.8	5	20	7	27
23:15–23:45	5.9	10	28	6	34
23:45–00:15	6.2	5	35	8	43
00:25–01:00	6.0	0	33	7	40
Totals	—	—	131	34	165

contemplated only if many time markers are recorded while the observations are being made. Ultimately, what you want to produce is a data table similar to Table 6.

The exact form of the data table is not critical. You may certainly design your own, but various organizations, such as the British Astronomical Association, the International Meteor Organization and the American Meteor Society, do have specific data table formats. With the data arranged as in Table 6, only the raw meteor count data have been collated. Next, the raw data will require some 'correcting' in order to provide a better estimate of the total shower activity.

Cloud Cover Correction

The cloud correction term, F, is evaluated in such a way that the greater the cloud cover, the greater the amount of correction is applied. That is, the more cloud cover there is, the greater the number of meteors that will on average be 'missing' from the counting statistics. The correction term is written as

$$F = \frac{1}{1-f} \quad (3)$$

where f is the estimated cloud cover during each of the observing time intervals. From Table 6 we see, for example, that in the first time interval, 22:15 to 22:45, $f = 0.0$, and the cloud correction term is thus $F = 1.0$. During the second time interval, however, $f = 0.05$ (i.e. 5 per cent), and consequently the cloud correction term is $F = 1/(1 - 0.05) = 1.05$. In the third time interval, from 23:15 to 23:45, $f = 0.1$ and thus $F = 1.11$. One can see from equation (3) that as f approaches 100 per cent so the correction F becomes larger and larger. For this reason, once f has become greater than about 30 to 40 per cent, you might as well stop collecting data – the correction term for such cloud conditions becomes unreasonably large, and the results are too unreliable.

The Population Index

Once the numbers of meteors visible during the observing time intervals have been collated, you should next determine the population index, r. This term is evaluated from the magnitudes of the meteors seen, and is defined as the ratio of the number of meteors $N(m + 1)$ of magnitude $m + 1$ to the number $N(m)$ of meteors of magnitude m:

$$r = \frac{N(m + 1)}{N(m)} \quad (4)$$

Table 7
A magnitude distribution table: values of N(m) for magnitudes from –4 to 6, for observations of Geminid and sporadic meteors (the first row is for mag. –4 to –3, the second for –3 to –2, etc.). If a meteor of magnitude –4 or brighter is observed, then a fireball report should also be made.

Magnitude	**N(m)**	
	Geminids	**Sporadics**
–4	0	0
–3	1	0
–2	0	0
–1	1	0
0	3	1
1	8	2
2	20	5
3	47	7
4	30	8
5	15	7
6	6	4
Totals	131	34

The population index will not necessarily be constant over the entire range of magnitudes of the observed shower meteors, but in practice it is assumed to be so. From its very definition, the population index is a measure of the relative meteor richness of the shower: r-values of 2 or less indicate a shower relatively rich in bright meteors, whereas r-values greater than about 3 correspond to a shower relatively rich in very faint meteors. Sporadic meteors are found to have a population index of about 3.0. The annual meteor showers have population indices that fall in the range $2.0 < r < 3.5$. The Lyrid meteor shower, for example, has a population index of 2.1, and is relatively rich in bright meteors. December's Ursid meteor shower, on the other hand, has a population index of 3.0, and is relatively rich in faint meteors.

To evaluate the population index you first have to determine the magnitude distribution of the meteors observed, for which you will need to compile a second table from your observing records. Table 7 shows one suggested layout. The numbers are for the same Geminid observing session as in Table 6, and consequently the total number of shower and sporadic meteors should be the same.

The determination of the population index from a magnitude data set requires a certain amount of pre-judgement and finessing. For example, we can see from Table 7 that although we would expect to see more meteors at fainter magnitudes, this is apparently not the case. Indeed, the data indicate a maximum count at magnitude +3, and a decline in the number of fainter meteors. This effect is mostly a result of the probability of perception discussed above – most of the faint meteors are simply missed by the observer. This condition indicates that the

Table 8
The variation of the population index (r-value) with magnitude for the example of the Geminid observing session. The r-value is determined from equation (4) with values for N(m) taken from Table 7.

Magnitude	**r-value**
–4	—
–3	0
–2	0
–1	3
0	2.7
1	2.5
2	2.4
3	0.6
4	0.5
5	0.4
6	—

population index should be estimated only for meteors brighter than the 'turnover' magnitude (magnitude +3 in the case of our sample). The r-value for each magnitude bin in Table 7 is given in Table 8.

The dramatic change in the r-value at magnitude +3 seen in Table 8 is related to the undersampling of faint meteors (the probability of perception effect) and is not 'real'. The overall population index for the Geminid shower is therefore calculated from the average r-values for magnitudes above +3. In our example, this gives a population index of r = 2.7.

Since the population index can vary with time, it should be calculated, if the data allow, as frequently as possible. Certainly, it is reasonable to determine the population index for every sequential set of 50 to 75 meteors recorded.

The Zenithal Hourly Rate

The activity of a meteor shower is described in terms of the number of meteors seen per hour. This number, however, is not the actual number that any given individual observer will record. The activity of a meteor shower is described in terms of a standardized zenithal hourly rate (ZHR). The ZHR is defined as the number of meteors that would be seen per hour, under perfect viewing conditions (i.e. with a limiting magnitude of +6.5 and zero cloud cover), if the shower radiant were at the observer's zenith. The ZHR is derived from the actual number, N, of meteors seen over a time interval T for which the population index r, limiting magnitude LM and cloud correction factor F have been determined:

$$ZHR = F \frac{N}{T} \frac{1}{\sin e} r^{6.5-LM} \quad (5)$$

The sin e term is an additional correction factor for the elevation, e, of the radiant above the observer's horizon. When the radiant is close to the horizon (i.e. when e is less than 10° to 15°), then virtually no meteors, unless they are of near-fireball status, will be seen and consequently a high correction term is required. If the radiant is at the zenith, then e = 90° (and sin 90° = 1.0), so no correction for elevation is applied. If the radiant has an elevation of less than 10° to 15°, it is not worth beginning a meteor count. Wait an hour or so for the radiant to move higher into the sky, and then begin.

The variation in the ZHR for our Geminid observing session can be derived from equation (5) and the correction terms derived in the tables above. The only additional information required is the radiant elevation, which is given in Table 9. The radiant elevation for the shower you are monitoring can be evaluated from any computer planetarium program.

Table 9
The deduced ZHR variation for the example of the Geminid observing session. A population index of r = 2.7 has been used in the calculation.

Interval (UT)	T (h)	LM	$r^{6.5-LM}$	F	N	e	ZHR
22:15–22:45	0.50	6.1	1.49	1.00	15	77°	45.9
22:45–23:15	0.50	5.8	2.00	1.05	20	83°	84.6
23:15–23:45	0.50	5.9	1.81	1.11	28	88°	112.5
23:45–00:15	0.50	6.2	1.35	1.05	35	83°	100.0
00:25–01:00	0.58	6.0	1.64	1.0	33	77°	95.8

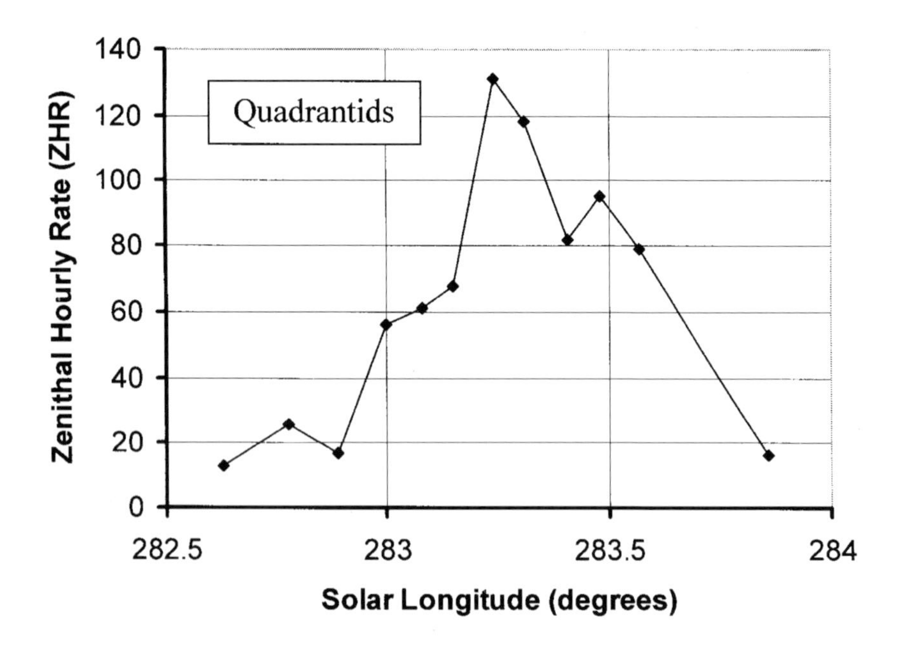

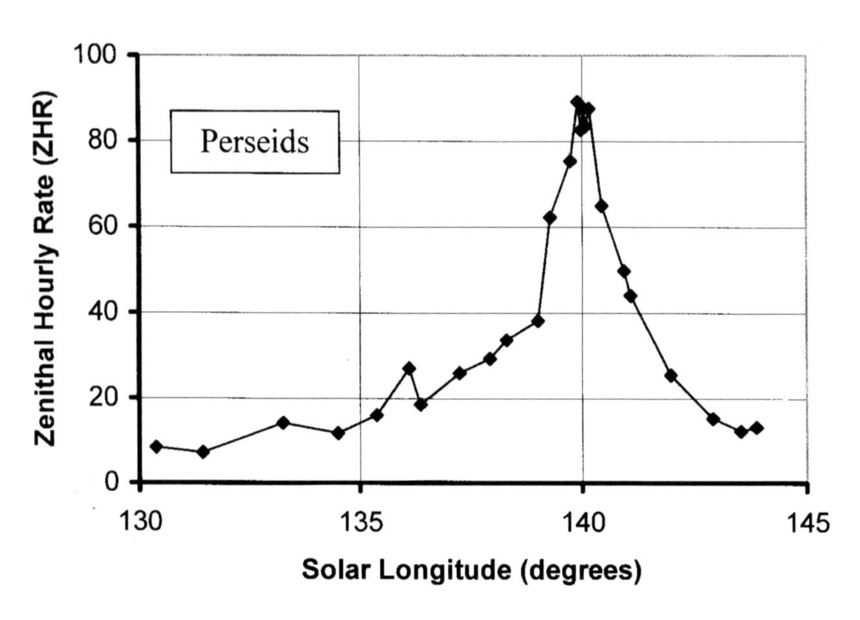

***Fig. 82** The activity profiles for the 2001 Quadrantid and 2002 Perseid meteor showers. Note that the solar longitude divisions on the x-axis are 0.5° (~12 hours) for the Quadrantids and 5° (~5 days) for the Perseids. (Data from the IMO archive)*

The Activity Profile

The activity profile of a meteor shower is expressed in terms of how the ZHR varies with time. In general, the activity profile of a given shower is constant from one year to the next, but the activity profile will vary significantly

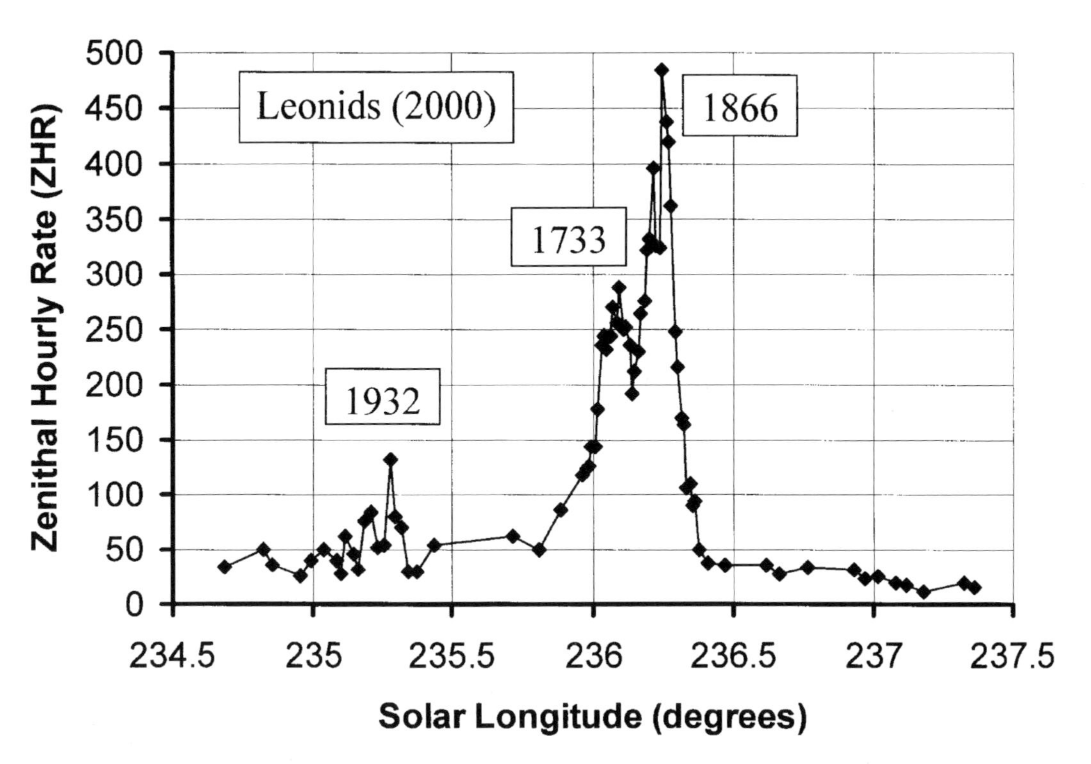

Fig. 83 *During the 2000 Leonid meteor storm it was found that the Earth encountered dense streamlets of meteoroids ejected from the parent comet, Tempel–Tuttle, in 1733, 1866 and 1932. The time the Earth took to cross each streamlet was about 2½ hours (corresponding to about 0.1° of solar longitude). This crossing time corresponds to the streamlets being about 270,000km wide, given that the Earth has an orbital speed of 30km/s. (Data from the IMO archive)*

from one shower to another. In most cases a single observer will not be able to observe a shower for long enough to determine a complete activity profile. Clearly, if a shower is active for several days, no single observer can observe its activity during the daytime or when the shower's radiant is below their horizon. The construction of a detailed activity profile is very much a project requiring a group effort from observers stationed around the entire globe. It is for this very reason that submitting your observational results to societies such as IMO, the BAA or the AMS is of great importance, as they can be of scientific value.

Figure 82 shows the activity profiles for the 2001 Quadrantid and 2002 Perseid meteor showers. It can be seen from the figure that the Quadrantid meteor shower is active for just a few tens of hours, whereas the Perseids are active for many weeks.

Although it might be thought that the activity profiles of the major annual meteor showers have been well determined, this is not the case. As described below, meteor showers occasionally undergo outbursts and storms, during which the hourly rate of meteors is significantly higher than average. In addition, some meteor showers show smaller-scale variations in their activity profile that last for perhaps an hour or so, and which vary from one year to the next (*see* Figure 83). The identification and monitoring of these transient features is one of the most important and useful studies that amateur meteor observers can perform.

Plotting a Shower's Radiant

Determining the positions of shower radiants was once a major preoccupation of meteor observers. It is less popular in the modern era, but it is still an important part of meteor astronomy. If you plan to hunt for new radiants – that is, for new meteor showers – you must first become familiar with the heavens. A good knowledge of the stars and constellations is vital if you are to make any useful headway in mapping meteor trails.

The equipment required for mapping a radiant is identical to that used for standard meteor counting. You will most definitely need photocopies of selected starfields from a good star atlas, such as the *Atlas Brno 2000.0*. If no specific shower is active on the night you plan to make a set of observations, then in principle you can look anywhere in the sky for meteors. It is probably best to start by selecting a few distinct starfields and constellations that you are familiar with and direct you attention towards them.

Record the details of your observing session: the location, the date and time, the region of sky being observed, the sky conditions and the limiting magnitude. Once you have in your mind's eye a clear correspondence between what you are seeing on the sky and what is printed on your sky charts, you can begin recording your observations. When you spot a meteor, the key response is to relax – don't rush to draw the meteor's trail on the sky chart. 'Savour' the image: make sure that you have fixed in your mind the points where the meteor trail began and ended in relation to the stars. Once this is clear, carefully place the meteor trail on your star chart. Number the trail and add an arrow to indicate its direction of motion. Then record the remaining information about the meteor: its maximum brightness, duration and colour, and the presence (or not) of a train. A final quantity to note for each meteor witnessed is an estimate of your observing accuracy. A comment such as 'high', 'medium' or 'low' will give some indication of how certain you are of the path and characteristics of the meteor. Only when you are satisfied that the meteor has been clearly plotted and the data have been correctly recorded should you go back to monitoring the starfield.

By the end of a good observing session, lasting perhaps several hours, you may have ten to twenty meteor trails on your star chart. The analysis stage now follows, and an extra complication arises. You will have plotted on a flat piece of paper the meteor trails you observed crossing the celestial sphere. Just as with maps of the Earth, there is an inherent distortion in the lines on any sky chart. That doesn't stop the charts being useful, of course, and the actual distortions produced by projecting a three-dimensional surface onto a two-dimensional one are mathematically determined. A familiar standard projection for terrestrial maps is the so-called Mercator projection, which projects the surface of a sphere onto a cylindrical surface. Why all this is a problem at the radiant analysis stage is that on an ordinary star chart meteor trails should strictly be plotted as arcs of great circles and not as straight lines. This is why it is most important to get an accurate fix on the exact beginning and end points of the trail since it is the right ascension and declination coordinates of these two points (read from your sky atlas plots) that will be used in the eventual analysis. Meteor trails may be plotted as straight lines only if a star chart based upon what is called a gnomic projection has been used. This is not the place to go into the detailed theory behind the production of gnomic charts, but for those who wish to follow the procedure, information is provided in various IMO publications and on a number of internet sites. However, rather than attempting to construct your own gnomic charts, you can use a computer program developed by Rainer Arlt called *Radiant*, downloadable from http://www.imo.net/software/. The input that the program requires is the right ascension and declination of the beginning and end points, and an estimate of angular speed in degrees per second, for each meteor observed. The program performs all the required coordinate transformations and automatically finds and identifies any possible radiants.

The identification of a new meteor shower requires more than finding a previously unknown radiant position. It has to be established that the new radiant has not come about by the chance orientation of several sporadic meteor trails. By their very nature, sporadic meteors travel in random directions across the sky, but there is always the possibility that four or five sporadic meteor trails can be traced back to what appears to be a radiant point. For this reason, the criterion for the identification of a new shower requires that observations be made over many years by different observers.

Telescopic Observing

Meteors fainter than magnitude of +6.5, the limit for the unaided human eye, can be observed through small telescopes and binoculars. The formula for calculating the limiting magnitude of a telescope (or binoculars) in terms of its aperture was given in Chapter 1. As one would expect, the formula indicates that the larger the telescope's aperture, the greater its limiting magnitude (i.e. the fainter the stars that can be seen). Now, while there are more faint meteors than bright ones (expressed by the population index as described above), this does not necessarily mean that more meteors will be seen though a telescope. This is mainly because the area of sky that can be seen through a telescope is much smaller than the area that can be monitored by a naked-eye observer. Typically, telescopic observers tend to see just a few meteors per hour. While the inherently low meteor detection rates make telescopic observing rather tedious (according to one's tastes), the limited field of view is beneficial in that if a meteor is seen then its path can be plotted with high accuracy.

When Stars Fall Like Rain

Imagine the scene: 'looking up to the sky we saw from time to time exhalations or stars, which soon went out, but without noise... All night it rained stars but we saw none fall to the ground.' This was the scene witnessed by rancher Eulogio Mijares in Zacatecas County, Mexico, on the night of 27 November 1885. The meteor storm that Mijares, and many others across the Americas, experienced was produced the Andromedid stream associated with comet Biela (now lost). At its maximum intensity it is estimated that something like 6,000 meteors per hour were seen. The sight of so many meteors must have been awe-inspiring. I will always remember the vivid display of the 2001 Leonid meteor storm I watched from Southern Saskatchewan, Canada, when in less than an hour over 200 bright fireballs were witnessed. Once seen, a meteor storm is never forgotten, and history abounds with accounts of stars falling like snow, or swarming like a cloud of locusts.

Storms and Outbursts

There is no formal definition of what marks the beginning of a meteor **storm**, but as with many complicated things, they are obvious once seen. Some have suggested that reaching a ZHR in excess of 1,000 meteors per hour is a good criterion for the onset of meteor storm conditions, while others have suggested a ZHR of 3,600, corresponding to one meteor per second. Less dramatic but more common than storms are **outbursts**, a term used for sustained enhancements of activity above normal meteor rates. Typically, the 'outburst' designation indicates that meteor rates some two to ten times greater than normal have been recorded on the night of the shower's maximum.

However, irrespective of how one might try to distinguish an outburst from a storm, the key point is that meteor counts should be continually gathered. Careful year-to-year monitoring of all meteor showers is required, since the normal, non-storm/non-outburst meteor rates are the standard by which outbursts and storms are ultimately measured.

Meteor Storms Through History

Spectacular meteor storms have been recorded throughout documented history. While most of the historical storm accounts cannot be easily evaluated in the sense of deriving a reasonably accurate ZHR, it does appear that some of the historical storms may well have produced meteor rates in excess of 100,000 per hour. Among the presently active meteor showers, the Lyrids, Leonids and Draconids have all historically produced storms and distinct outbursts. As mentioned above, the now inactive Andromedid meteor shower is known to have experienced several storms and outbursts. The annual Perseid, Orionid and Ursid meteor showers have also undergone distinct outbursts, but have not apparently produced any historical meteor storms.

Many meteor showers have been observed to undergo outbursts when their parent comets are close to perihelion. This has happened with the Ursids, Perseids and Draconids, for example. The Leonids are perhaps the most extreme in this respect in that the shower can produce storms or outbursts at 33-year intervals, coinciding with the orbital period of the parent comet. The Draconids are generally considered to be an 'erratic' shower in that the annual activity rate is highly variable. The shower did undergo distinct storm activity in both 1933 and 1946, when the parent comet, Giacobini–Zinner, was at perihelion. At other returns of the comet, however, no enhanced activity was seen. The Lyrid meteor shower has undergone a number of outbursts (e.g. in 1922, 1934, 1946 and 1982) and at least two storms (in 1122 and 1803). The Lyrid outbursts are not actually associated with the perihelion return of the stream's parent comet – the long-period comet, comet Thatcher, which has an orbital period of 415 years – and interestingly they appear to be separated by multiples of 12 years. This may be the result of stream 'sculpting' brought about by gravitational interactions with the planet Jupiter, which has an orbital period of 11.9 years. The Taurid meteor shower is also known to undergo semi-periodic outburst activity, and these outbursts appear to be governed by a complex gravitational interaction with Jupiter as the result of which higher Taurid meteor rates are seen at intervals of 3, 4 or 7 years.

The Andromedid meteor storms of November 1872 and 1885 are interesting in that they were caused by the break-up of the stream's parent comet, comet Biela. The comet appeared to break into two components during its 1846 perihelion passage and was actually seen as two distinct comets at its 1852 return. The cometary fragments have not been seen since 1852, and the associated Andromedid shower is no longer active. It has been speculated that comet Biela actually broke apart when it passed through a dense part of the Leonid meteoroid stream in 1832. Computer studies of the comet's orbital evolution indicate that there is a small possibility that the Earth may once again encounter meteoroids from comet Biela early in the twenty-second century.

A study I carried out with Peter Brown of the University of Western Ontario, published in 1995, found that the incidence of meteor storms since the end of the first millennium AD has rarely been less than two per century. Some centuries have fared very much better than

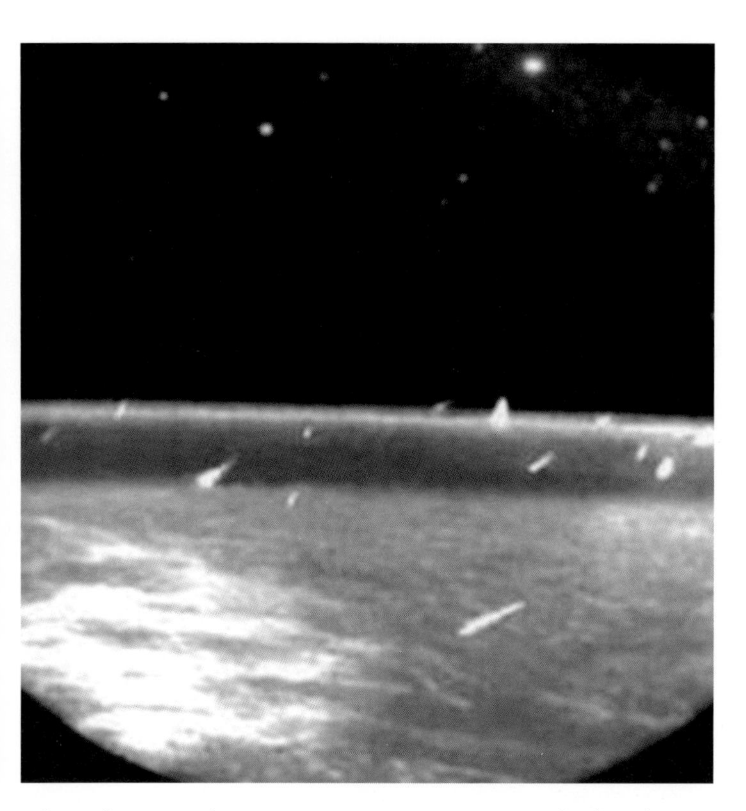

Fig. 84 *The Leonids from space. This composite image shows Leonid meteors ablating in the Earth's upper atmosphere, as recorded by cameras on board the Midcourse Space Experiment satellite in November 1997. Stars in the constellation Aries are visible in the upper half of the image, and Moon's glare is visible on cloud-tops to the lower left. (Image courtesy P. Jenniskens, SETI Institute, NASA ARC)*

others: for example, seven meteor storms were recorded during the nineteenth century, and six meteor storms were witnessed and recorded during the twentieth century.

During the past two centuries the Leonid shower has dominated the meteor storm count, but it is expected to dwindle during the twenty-first century. Of the showers that have produced storms during the last century, the Draconids are perhaps the most likely to deliver a storm during the present one. After the last perihelion return of comet Giacobini–Zinner, in November 1998, the Draconid shower underwent a strong outburst, with a visual meteor rate of some 100 per hour being recorded at the time of the shower's maximum activity. The comet next rounds perihelion in 2011 and again in 2018. These may well be good years to be out observing on the night of 8 October, the night of the shower's maximum. The 2018 return will be particularly interesting in that the comet will pass within 0.4AU of the Earth on the night of 11 September. Among the weak annual meteor showers, it has been predicted that the Aurigids might undergo a storm or outburst on 1 September 2007. And it has been suggested that the Tau Herculids (associated with comet Schwassmann–Wachmann 3) might undergo a storm or outburst on 31 May 2022.

Odd Meteors

Observers occasionally report seeing odd-looking meteor trails, and some are indeed phenomena of great curiosity. Some of the odd trails reported are apparently wavy or sinusoidal, some are nebulous and fuzzy, and others appear to move along helical paths. Some trails appear to curve in a continuous arc while others are 'bent' in the sense that the meteor appears to have abruptly changed its direction of motion.

Fig. 85 *Comet Giacobini–Zinner, the parent comet of the Draconid meteor shower. This short-period comet is not especially dust-rich, and in this image only a Type I plasma tail is visible. (Image by N. A. Sharp, NOAO)*

Fig. 86 *In 1995 the nucleus of comet Schwassmann–Wachmann 3 was observed to fragment. In the three upper images, obtained with the 3.5m New Technology Telescope at the La Silla Observatory in Chile, three distinct nuclei can be seen. The lower image was taken with the 3.6m telescope at La Silla, and shows an infrared image of the cometary nuclei. It has been predicted that smaller meteoroid fragments released during the break-up will find their way to the Earth towards the end of May 2022. (Image courtesy the European Southern Observatory)*

A number of years ago I made a study of the reporting frequency of odd-looking meteor trails. I found that experienced observers – those who had gathered many hundreds of hours of observations – reported that one meteor trail in about two hundred had some strange or noteworthy characteristic. The reporting statistics further suggest that about 50 per cent of the odd-looking trails are continuously curved, 40 per cent are wavy or sinusoidal and the remaining 10 per cent show abruptly bending trails or strange fragmentation characteristics.

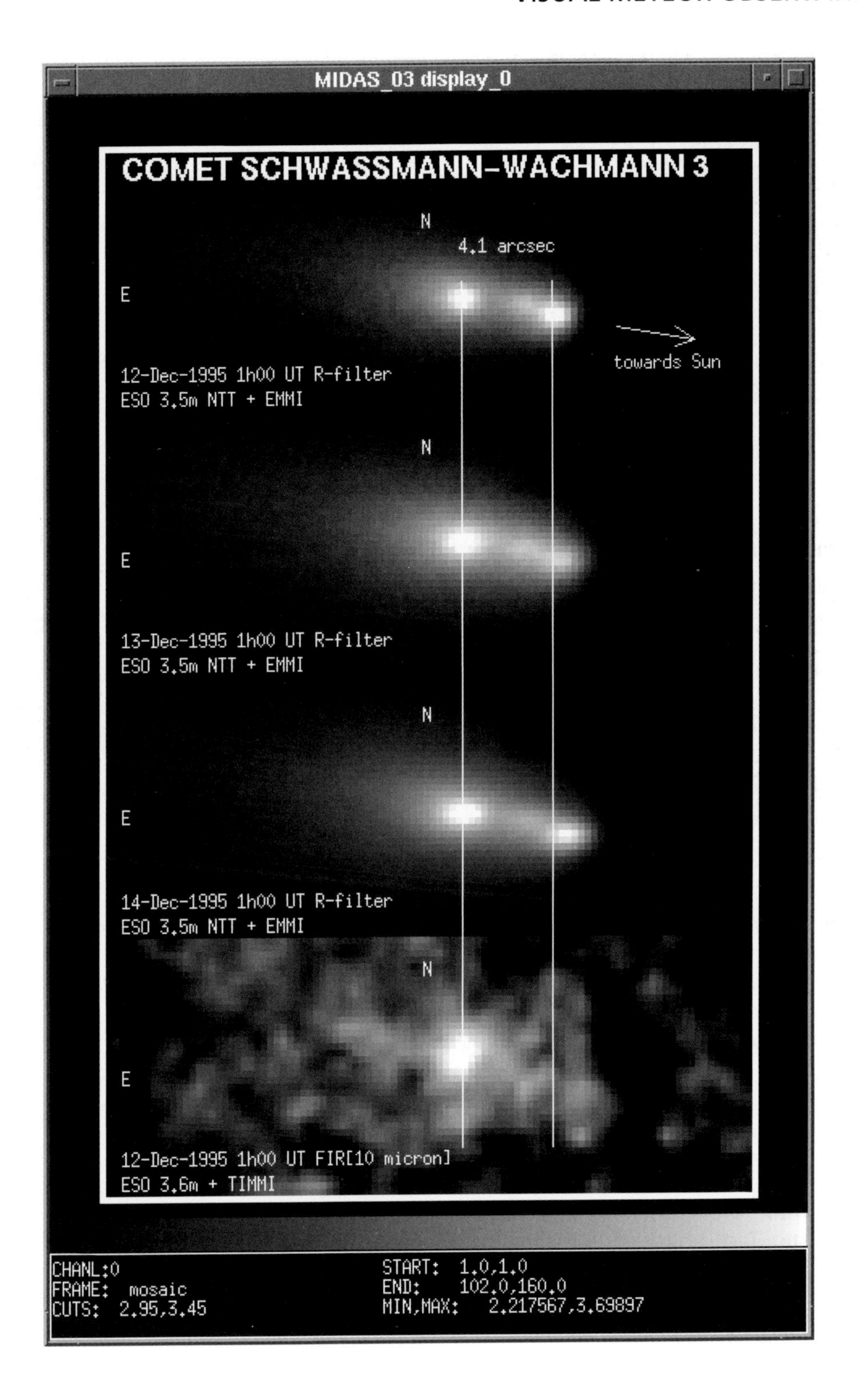
MIDAS_03 display_0
COMET SCHWASSMANN-WACHMANN 3
N
4.1 arcsec
E
towards Sun
12-Dec-1995 1h00 UT R-filter
ESO 3.5m NTT + EMMI
N
E
13-Dec-1995 1h00 UT R-filter
ESO 3.5m NTT + EMMI
N
E
14-Dec-1995 1h00 UT R-filter
ESO 3.5m NTT + EMMI
N
E
12-Dec-1995 1h00 UT FIR[10 micron]
ESO 3.6m + TIMMI
CHANL:0
FRAME: mosaic
CUTS: 2.95,3.45
START: 1.0,1.0
END: 102.0,160.0
MIN,MAX: 2.217567,3.69897

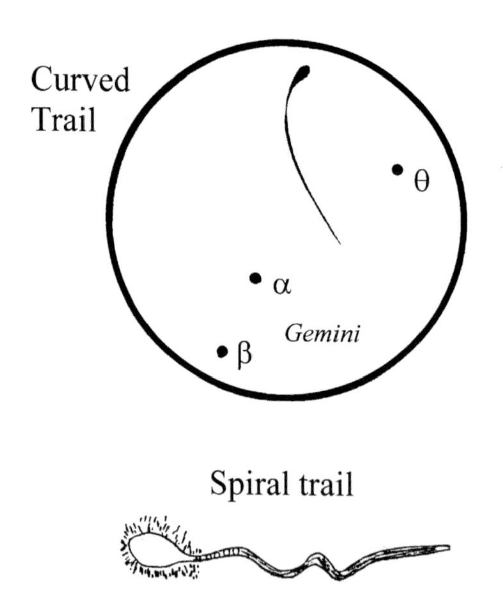

Fig. 87 *The two main types of odd trail reported are the curving trail and the spiral or wavy trail. The curved trail shown here was observed and drawn by famed meteor observer William Denning on 25 December 1886 as it passed through the constellation Gemini. Alexander Herschel observed the spiral trail shown here on 18 July 1900 and described it as 'the queerest meteor I ever saw'.*

If you chance to observe a meteor trail that looks genuinely odd, then make as many notes on its appearance as you possibly can. Draw a sketch of the trail while it is fresh in your memory. Record the time at which the event occurred and make a note of the viewing directions. Estimate the magnitude, duration and angular speed, and assess any potential shower membership. Since such events are so rarely reported, it is always worth sending your observations to a journal such as *WGN*, the bimonthly publication of the International Meteor Organization. Most of the odd meteor trails reported in the past appear to have been associated with sporadic meteors, but since the physical mechanisms responsible for many of the reported trail shapes are entirely unknown, there is no reason to say that some shower meteors cannot show odd trails. There is always the possibility that at least some odd-looking meteor trails are an illusion, produced by some currently unrecognized physiological effect. This possibility could be investigated by trying to gather data on viewing circumstances, so make a note of how you saw the meteor: whether you turned you head to see it, whether it was seen at the end of a long observing session, and so on. A few photographs of apparently sinusoidal meteor trails have been published over the years, but a real scientific coup would be to capture a video sequence of such a trail.

One of the interesting and presently unexplained surprises that has emerged from the wealth of Leonid video observations gathered during the past few years was that perhaps one in a hundred Leonids appeared to be nebulous, with the light being emitted from an extended region (*see* Figure 88). Typically, a meteor's light comes from a cylindrical region perhaps 5 to 15m across, but these nebulous Leonids appeared to produce light from a region several hundred metres across. The physical mechanism behind the nebulous appearance is unknown, and the effect is certainly worth looking for in video captures of other meteor showers.

Artificial Meteors

Not all the transient flashes that you might see in the sky are due to natural meteors – some are artificial. Since the launch of the Sputnik 1 spacecraft in 1957, an ever-increasing swarm of satellites, rocket bodies and general space 'junk' has grown around the Earth. Occasionally some of this debris falls back to the ground, interacting with the Earth's atmosphere to produce a meteor-like trail. Re-entering satellites are typically slow-moving

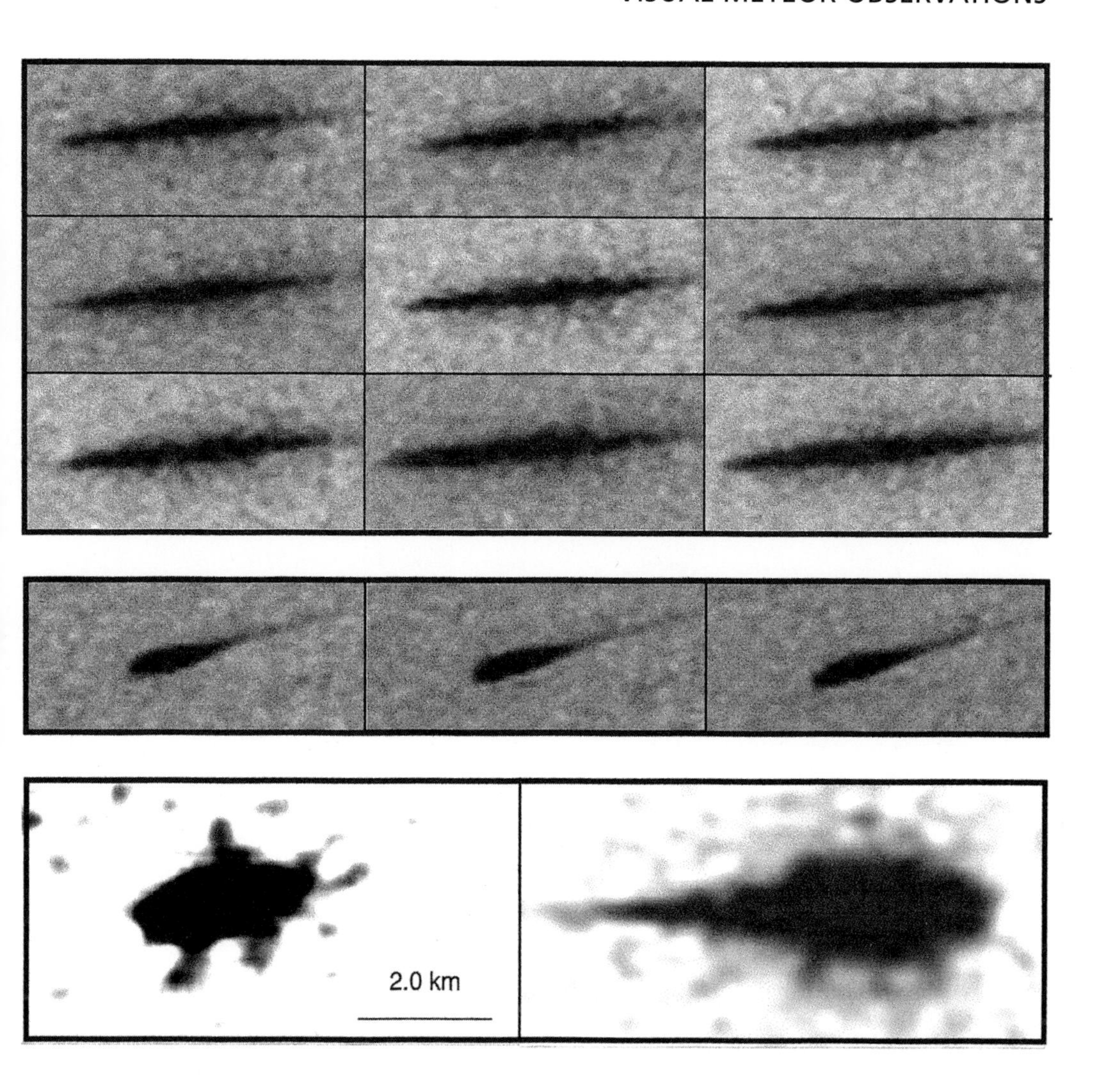

Fig. 88 *'Strange' Leonid meteors. The top set of images is a series of nine frames from a video sequence lasting just a few seconds. For this Leonid, a broad band of luminosity is seen transverse to the direction of the meteor's motion. An 'ordinary' Leonid meteor of comparable brightness is shown in the middle panel for comparison – it is much more compact in appearance. The lower panel shows a set of enlarged, single video frames of two Leonid meteors exhibiting jet-like features. The origins of the jets are completely unexplained at present. (Images courtesy Ian Murray, University of Regina)*

and trace out long trails across the sky. The Space Track website at www.space-track.org/ is the official outlet of the US Department of Defense satellite tracking initiative, and provides data on recent as well as forthcoming satellite re-entries.

***Fig. 89** The re-entry of the Mir space station on 23 March 2001. It broke into numerous fragments, each resembling a fireball as they descended through the Earth's atmosphere. Additional pictures of the Mir re-entry can be found at http://satobs.org/mir.html/. (Image courtesy Reuters)*

In recent years a new and distinct class of artificial 'meteor' has arisen in the form of sun-glints from the solar panels attached to satellites in low-Earth orbit. These artificial meteors are particularly useful for testing cameras and other recording equipment, since the satellite glints are essentially a slow meteor made to measure, and their times of occurrence and locations in the sky are predictable. The Iridium satellites, which are used for global voice and data communications, are one group of satellites that produce so-called Iridium flares. A number of websites provide detailed information on where and when to look in the sky for Iridium sun-glints. One of the most comprehensive set of web pages is that hosted by Heavens-Above, at www.heavens-above.com/. It is certainly worth checking to see whether any Iridium flares will be visible during your planed meteor observing session – they are certainly worth looking out for, and foreknowledge of such events will avoid any potential confusion with respect to the appearance of a real meteor or fireball.

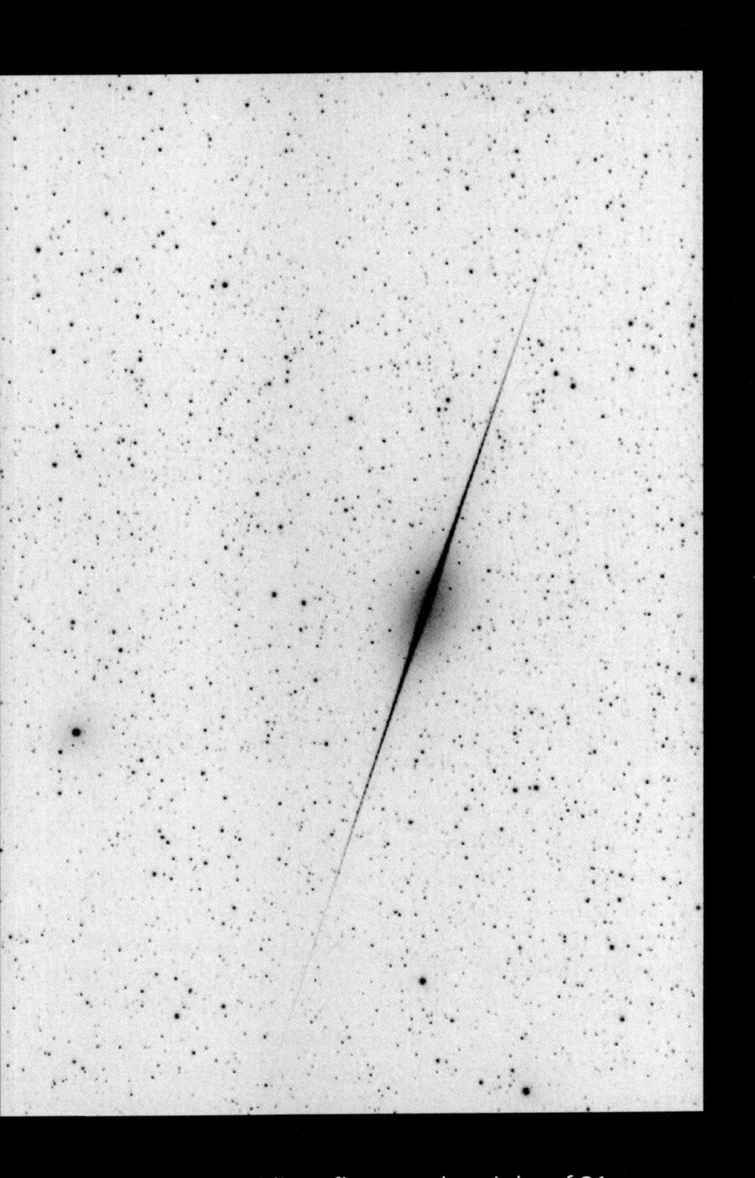

Fig. 90 *An Iridium flare on the night of 21 May 1998. Note the spindle-like nature of the trail. A 135mm, f/4 lens and Kodak 100 film were used to secure the image. (Image courtesy James Young, TMO/JPL/NASA)*

6 Photographic and Video Observations

As a highly mobile, low-light-sensitive, wide-angle imaging system, the human eye is a very efficient instrument – indeed, evolution has refined it to near-perfection for what it is normally required to do. Its main drawback when called upon to perform detailed astronomical work, however, is its relatively low resolution and its lack of positional accuracy (although the latter problem is partly related to an individual's observing skill and experience). It also suffers from the lack of an attached 'hard copy' recording system. If, therefore, you want to measure the more subtle characteristics of a meteor's trail, determine a meteoroid's speed and deceleration, or derive a meteoroid's orbit and atmospheric trajectory with any degree of

***Fig. 91** Left: an SLR camera with a standard 50mm focal length lens and lens cap attached. Right: an SLR camera with an all-sky (fish-eye) lens.*

precision, then photographic rather than eye observations are essential. A hundred years ago, in 1905, Professor Herbert Hall Turner eulogized that 'the wonderful exactness of the photographic record may perhaps best be characterized by saying that it has revealed the deficiencies of all our other astronomical apparatus – even the observer himself'. Canadian astronomer Peter M. Millman further commented, in 1934, that 'a good photograph of a meteor is worth, at the very least, several hundred visual plots'. For our purposes, however, the most important point about photography is that what the camera and film 'see' they record with great accuracy and they record it in a (semi-)permanent fashion.

Video observations, although they produce images of much lower spatial resolution, can complement photographic studies by enabling a greatly superior measure of time-varying quantities. As we shall see in this chapter, a video sequence can be separated into its individual frames, thereby revealing a sequence of images – a storyboard of change and movement. It is certain that the human observer will never become redundant in the field of meteor studies, but the human senses can certainly be extended by the use of photographic and video techniques.

Basic Photographic Techniques

Only a small financial investment is needed to begin photographing meteors. You will need a single-lens reflex (SLR) camera with a standard 50mm focal length lens, a sturdy (repeat, sturdy) tripod upon which to mount the camera, a lockable shutter release cable and fast film. The equipment does not have to be new or state of the art – in fact, as it will typically be operating in cold, damp or even frosty conditions, a second-hand camera body is probably preferable. However, one element that is worth investing as much money as possible in is the lens, which is critical to your success.

Much has been written about the best lens and camera combinations to use in meteor photography. Although there are many variables that can in principle come into play, in most cases there are just a few key ones that are likely to have an impact on how many meteors you might eventually photograph. Donald McKinley, who in 1961 wrote the highly readable but now somewhat dated book *Meteor Science and Engineering*, gave an empirical formula for estimating the efficiency E of a meteor camera system – although he offered no hint as to where the formula might have come from. McKinley maintained that the important quantities for estimating the efficiency are the focal length f of the lens being used, the area A of the sky being monitored, and the lens aperture a. The formula is

$$E = ka^2A/f$$

where k is a constant. Now, lenses are typically described in terms of the focal ratio (or f-number), which is the ratio of the focal length of the lens divided by the aperture. The smaller the aperture, the smaller the quantity of light transmitted to the film, and the larger the focal ratio. Perhaps confusingly, a small aperture corresponds to a large f-stop number, and a large aperture corresponds to a small f-stop number. The f-stop number is the number printed on the lens diaphragm ring. Typically it varies from 1.8 to 22, 22 indicating the smallest aperture setting and 1.8 the largest. In meteor photography the aim is the make the focal ratio as small as possible – and this is achieved by having the f-stop set to its smallest possible value.

The amount of sky area A covered by a camera system is related to the focal length of the lens and the size of the film being used. For standard 35mm film, each image is a rectangle 36mm long (L) and 24mm wide (W), and the sky area covered (in square degrees) is

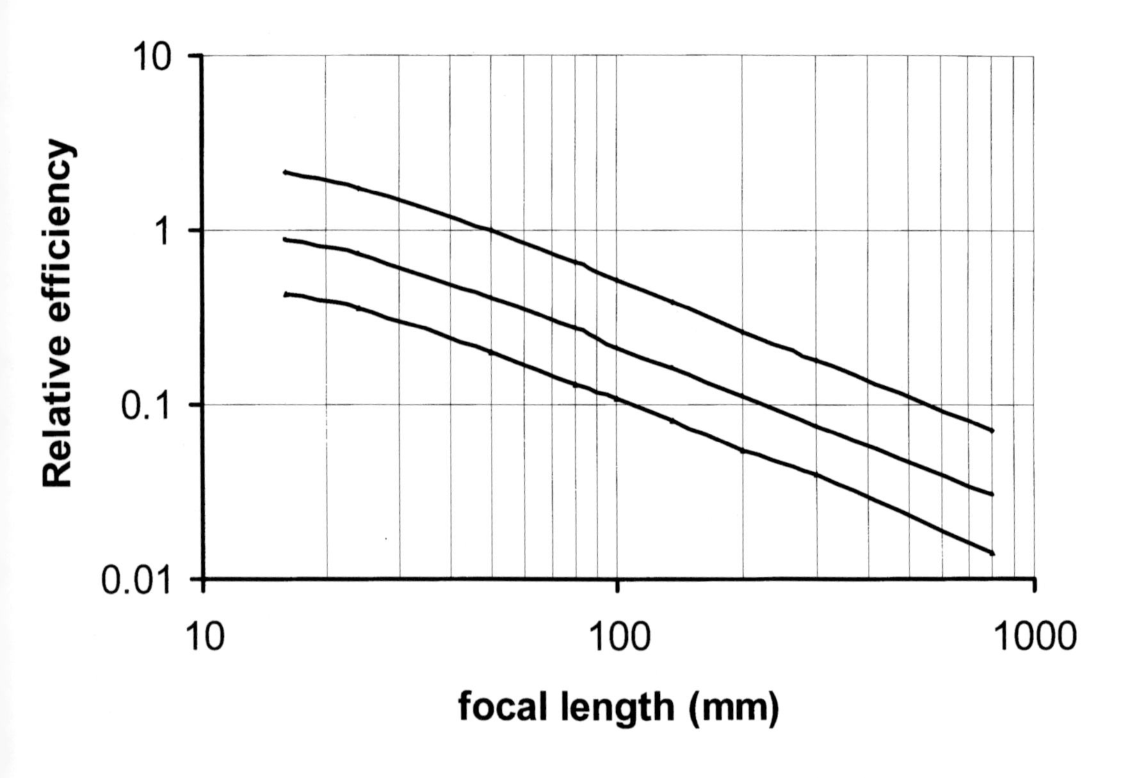

Fig. 92 *Relative lens efficiency for various combinations of focal length and f-stop number. The comparison for each curve is to a 50mm focal length lens with an f-stop setting of 1.8. The top line corresponds to an f-stop setting of 1.8, the middle curve to an f-stop setting of 2.8 and the bottom line to an f-stop setting of 4.*

$$A = 4(180/\pi)^2 \arctan(L/2f) \arctan(W/2f)$$

where f is the focal length in millimetres. Here 'arctan' is inverse or arc tangent function (scientific calculators have this as a standard button). For a 50mm focal length lens, the area of sky covered is thus found to be A = 1,069 square degrees – which corresponds to about 1/19 of the entire hemisphere of the sky. A fish-eye lens (*see below*) will cover the entire hemisphere of the sky and have a collecting area of 20,626 square degrees.

In terms of efficiency, as given by McKinley's formula, we can now evaluate the effect of varying the focal length and the f-stop setting. We will take a 50mm, f/1.8 lens as our standard for comparison, and assume that 35mm film is being used. Figure 92 shows the effect of varying the focal length and f-stop number. The efficiency is a trade-off between the sky coverage (which decreases with increasing focal length) and aperture (which increases with decreasing f-stop number).

McKinley argues in his book that the number of meteors captured per hour on film will vary in step with the efficiency. Figure 92 indicates that a 100mm, f/1.8 lens should capture about half as many meteors (for the same exposure time) as the standard 50mm, f/1.8 lens. In the same exposure time a 16mm, f/1.8 lens will capture about twice as many meteors as the standard lens. At f-stop settings of 2.8 and 4 the meteor capture efficiency of a 50mm lens will be reduced to 40 and 20 per cent of that for the standard lens. There are many additional factors that will modify these comparative figures, but the general message is clear: for scientifically useful results, one simply needs to use a standard 50mm focal length lens with as small an f-stop number as it can be set to. To improve upon the efficiency of the simple 50mm focal length lens system, one will have to either buy a shorter focal length lens, or invest in an expensive camera that takes larger format film. By moving to 120 square format film (with L = W = 55mm), for example, about three times the area of sky can be covered with a 50mm focal length lens than with 35mm film, with three times the potential for capturing meteors. The main drawback of using large format films is the very high cost.

With a standard 50mm, f/1.8 lens and standard ISO 400, 35mm film, one might hope to capture two to four meteor trails during a six-hour observing session when a rich meteor shower is active. When no major meteor shower is active, the number of meteors captured might be just one or two per six-hour session. If observations are being made outside the time of a major shower, then the camera can in principle be pointed in any azimuth direction. Just as with visual observations, the optimum elevation angle for the camera is 50° above the horizon. When a shower is active, the camera should be aimed in azimuth 180° away from the radiant, and elevated according to the radiant elevation as shown in Figure 93. Shower meteors that appear within a few tens of degrees of the radiant will have relatively low angular velocities, and a better chance of being captured on film.

Fig. 93 The optimum elevation angle for a camera (used with a standard 50mm lens) for different elevations of the radiant. (Data taken from the IMO Handbook for Visual Meteor Observers*)*

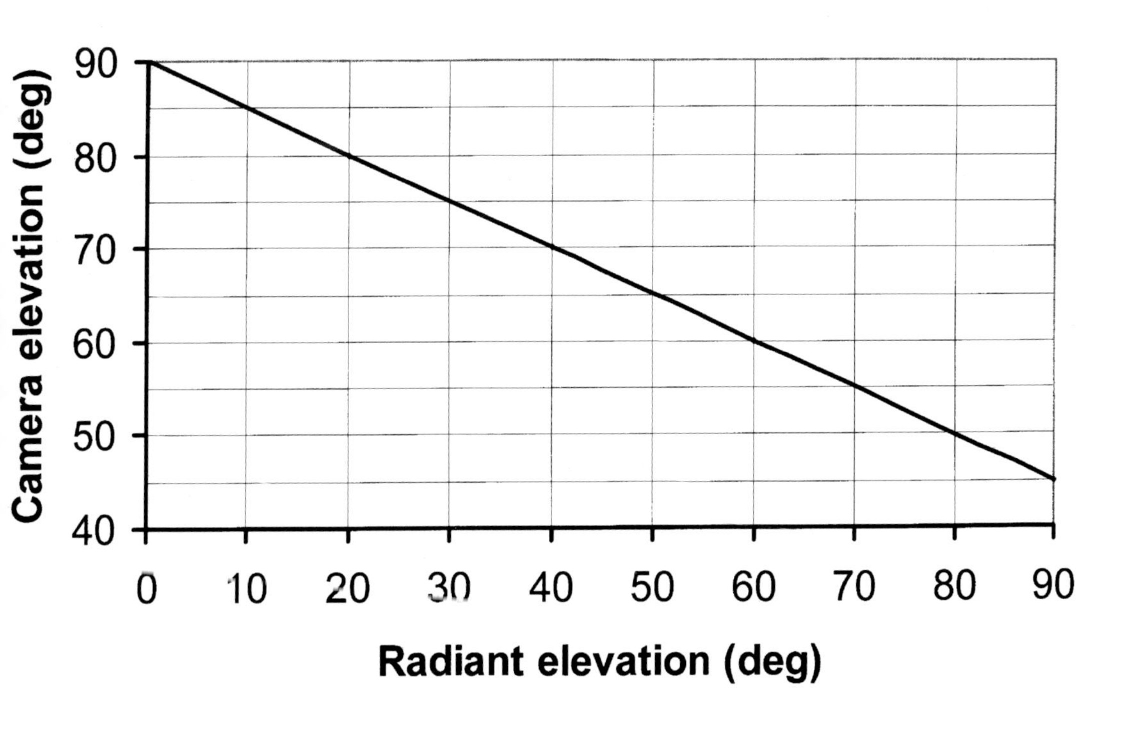

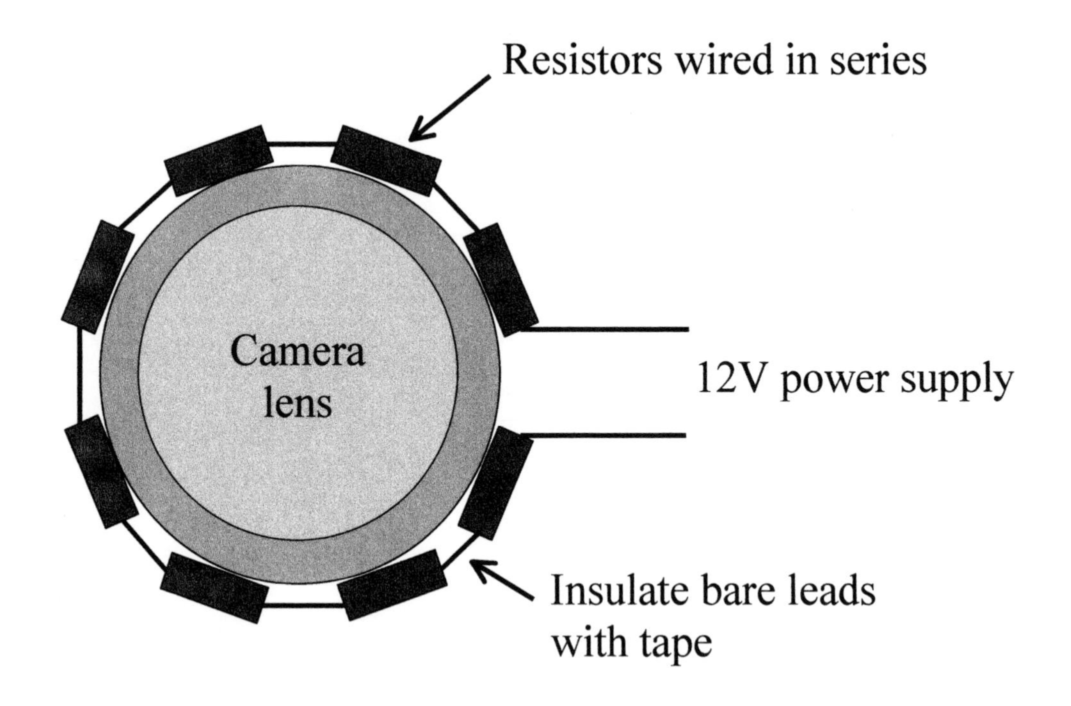

Exposure Times, Fogging and Dew Zapping

Fig. 94 *A basic resistor ring for heating a camera lens. Once the resistors have been soldered together and adjusted to fit the circumference of the camera lens, tape over the bare wires to prevent any short-circuiting from occurring. Fix the complete resistor ring around the lens with tape or with a wide elastic band.*

Just as with visual observing, meteor photography should be conducted from a dark location free of light pollution, and on nights when there is little or no moonlight. A location with minimal light interference is essential for long-exposure photography. Exposure times of 10 to 15 minutes are usual, but the time will vary according to the observing conditions, and only experimentation at your selected site will determine the best exposure time to maximize the number of meteors captured and minimize fogging and film costs. Fogging is an unavoidable hindrance in meteor photography and is caused by general background light pollution and the natural skyglow, too faint to be detected with the human eye.

The chances of capturing a meteor improve with increasing exposure time, but so too does the build-up of fogging from background light. One can minimize the fogging by using shorter exposure times, but this will result in more film being used. During the major annual meteor showers, shorter exposure times of perhaps 5 minutes can be used since the hourly rate of meteors is greater than when no shower is active. When you start using a new observing location, take a series of photographs with exposure times of 1, 2, 4, 8 and 16 minutes. Keep notes on the observing conditions, for example the Moon's phase and the direction in which the camera is pointing (note that towns and cities will produce localized sky-glows), and then examine the series of negatives for fogging and the limiting stellar magnitude. You are looking to find the longest exposure for which faint stars are clearly recorded and for which there is little or preferably no fogging.

On cold and frosty nights it is not uncommon for dew and eventually ice to form on the camera lens. Clearly this is not good for the camera, and, more importantly, it is detrimental to the quality and focus of any images that might be obtained. The build-up of dew and frost can be controlled, however, with a so-called dew zapper. A lens hood (normally used to shield the lens from direct sunlight) will help to minimize dew and frost build-up, but under extreme conditions a small electrical heating element is required. The temperature of the lens need only be very slightly higher than that of the surrounding air to prevent dew and frost from forming, so a low-voltage heater is all that is required. Electrical dew zappers can be brought commercially, or constructed at home from a few resistors wired together in series.

Typically, a heating power of about 3 watts is all that is needed to ensure that a camera lens doesn't become dewed over. The thermal power, P, dissipated by a ring of N resistors, of resistance R each, is given by $P = V^2/NR$, where V is the voltage of the power supply. If you run your dew zapper system from a 12-volt car battery, then the 3 watts of heating power can be generated by a string of identical resistors, wired in series, such that NR = 48 ohms. Resistors can be purchased from almost any

Fig. 95 *A bright Perseid fireball cuts across the constellation Ursa Minor. The Pole Star is at the lower right of the image and it can be seen that the star trails increase in length towards the left. The breaks in the fireball's trail are due to a rotating shutter being placed in front of the camera lens.*

Fig. 96 *This bright Perseid meteor was captured during a 15-minute guided camera exposure. Kodachrome 1000 slide film was used with a 30mm, f/2.8 lens. The bright star to the left of the meteor trail is Epsilon Eridani. (Image courtesy the European Southern Observatory)*

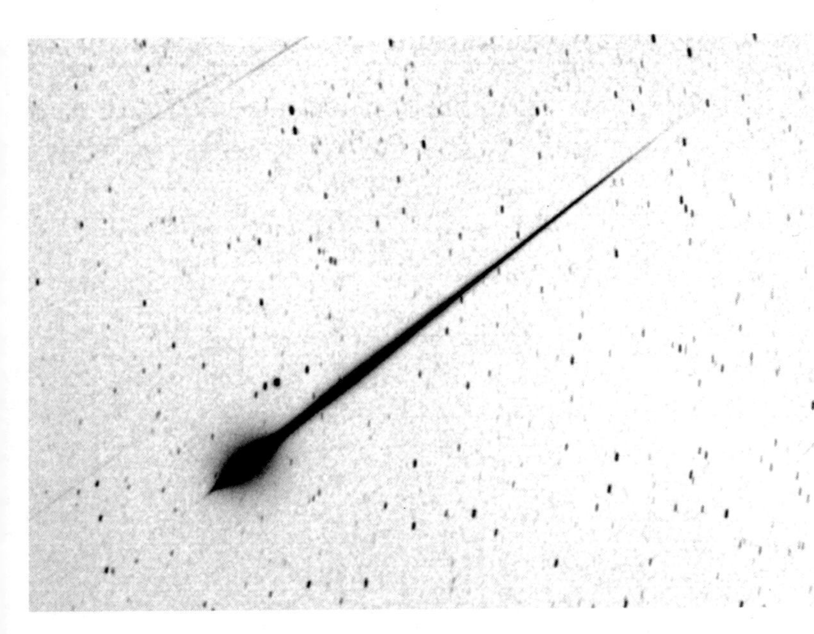

***Fig. 97** A bright Leonid fireball captured during a 2-minute unguided exposure during the 1966 Leonid storm. A 35mm, f/2 wide-angle lens and Kodak Plus-X (ISO 100) film were used to capture the image. Three fainter Leonid meteor trails are also visible. (Image courtesy James Young, TMO/JPL/NASA)*

electrical store (or via the internet). So, for example, if you buy a pack of resistors with individual resistances of 6 ohms, then a ring of eight resistors will provide the required total resistance. Figure 94 shows the arrangement of resistors in a basic dew zapping lens heater. Make sure that the resistor ring is large enough to fit around the rim of the camera lens that you will be heating.

To Track or Not to Track?

Camera tracking to compensate for the rotation of the Earth, so that the stars appear as point images on the negative and not as elongated trails, is not an essential requirement of meteor photography. If you are only interested in capturing images of meteor trails, then the camera can simply be attached to a fixed tripod and the stars allowed to drift through the field of view during the exposure. The lengths of star trails will increase with exposure time and will also vary with the declination, δ. Stars close to the north celestial pole (for which $\delta \sim 90°$) will have very short trails, whereas stars close to the celestial equator (for which $\delta \sim 0°$) will have long trails. Also, the longer the exposure time, the longer the star trail lengths. Figure 95 shows an example of a Perseid fireball image captured with a stationary, non-tracked camera.

When a camera is tracked to compensate for the Earth's rotation, the right ascension and declination of the meteor's beginning and end points are easily determined by comparing the photographic image with a star atlas. This is advantageous, for example, if one is trying to study the structure and the daily movement of a shower's radiant. Figure 96 shows a bright Perseid meteor trail captured during a tracked exposure. (It can be argued that tracked images are aesthetically more pleasing to the eye than non-tracked images, but beauty is always in the eye of the beholder.) An alternative to tracking is to take short exposures. This will result in much more 'wasted' film, in the sense that most images will be devoid of meteor trails, but during a meteor storm or outburst it may well produce good results. Figure 97 shows a very nice example of a Leonid fireball captured during a brief exposure – sometimes luck can be with you.

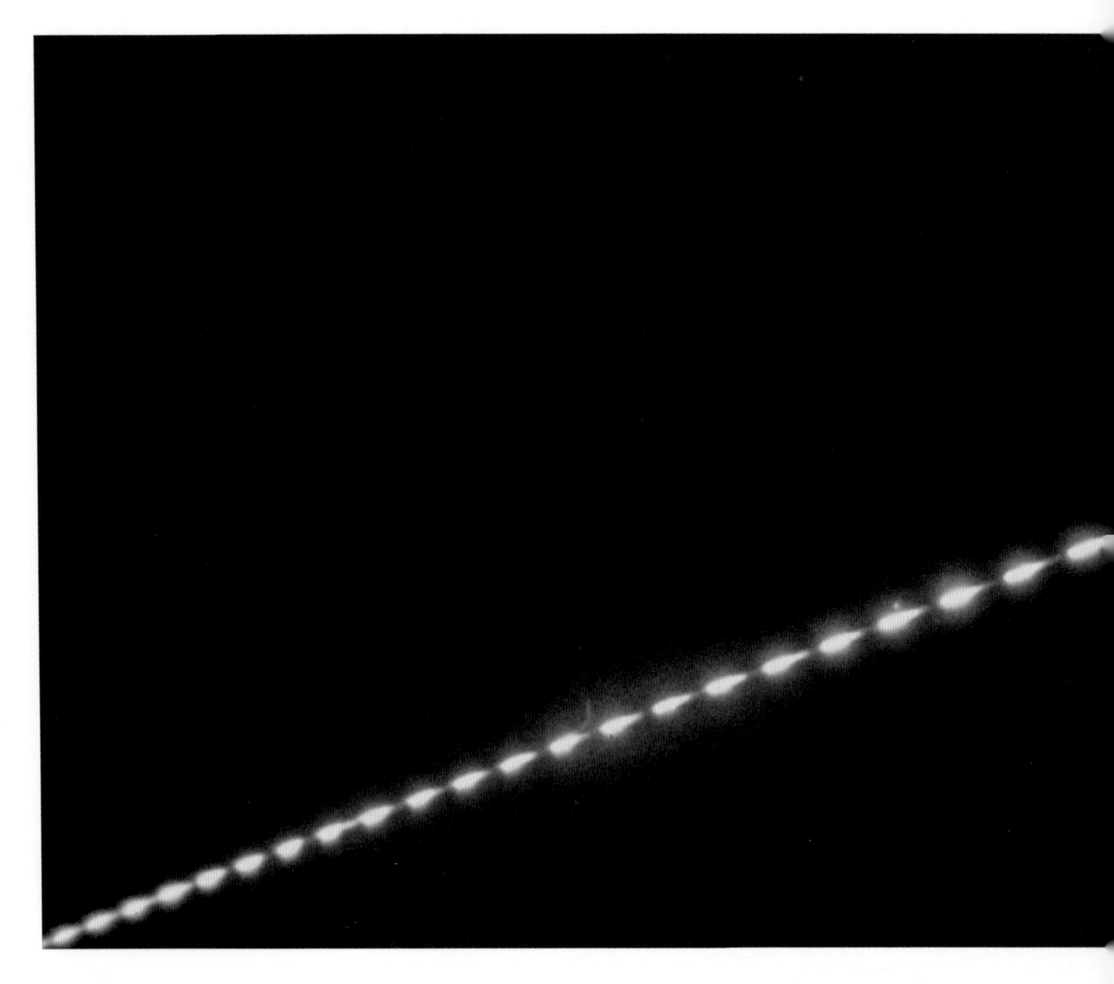

Film and Developing

The ability of a camera to capture meteors is determined partly by the quality of the lens being used and by its aperture (as discussed above), and partly by the sensitivity of the film being employed. The speed or light sensitivity of a film is rated on a scale established by the International Organization for Standardization (ISO). The higher the ISO number, the faster (i.e. the more sensitive) the film. Standard colour films typically have ISO ratings between 100 and 200, making them good for daylight photography, but not so good for meteor work. Indeed, it is worth noting that there is absolutely no advantage in using colour film in meteor photography. Films can be rated as high as ISO 3200, but ISO 400 is the typical film used by most meteor photographers. One of the drawbacks of using very fast film is the general graininess of the image, but since great image magnification is typically not required, this is not a major issue.

Ilford HP5 and Kodak Tri-X (both ISO 400) are two films that are commonly used in meteor photography and they have the advantage of being generally available from most photographic suppliers. Black and white film may be bought in standard 24- and 36-frame canisters, or can be purchased as a long roll from which film strips of any required

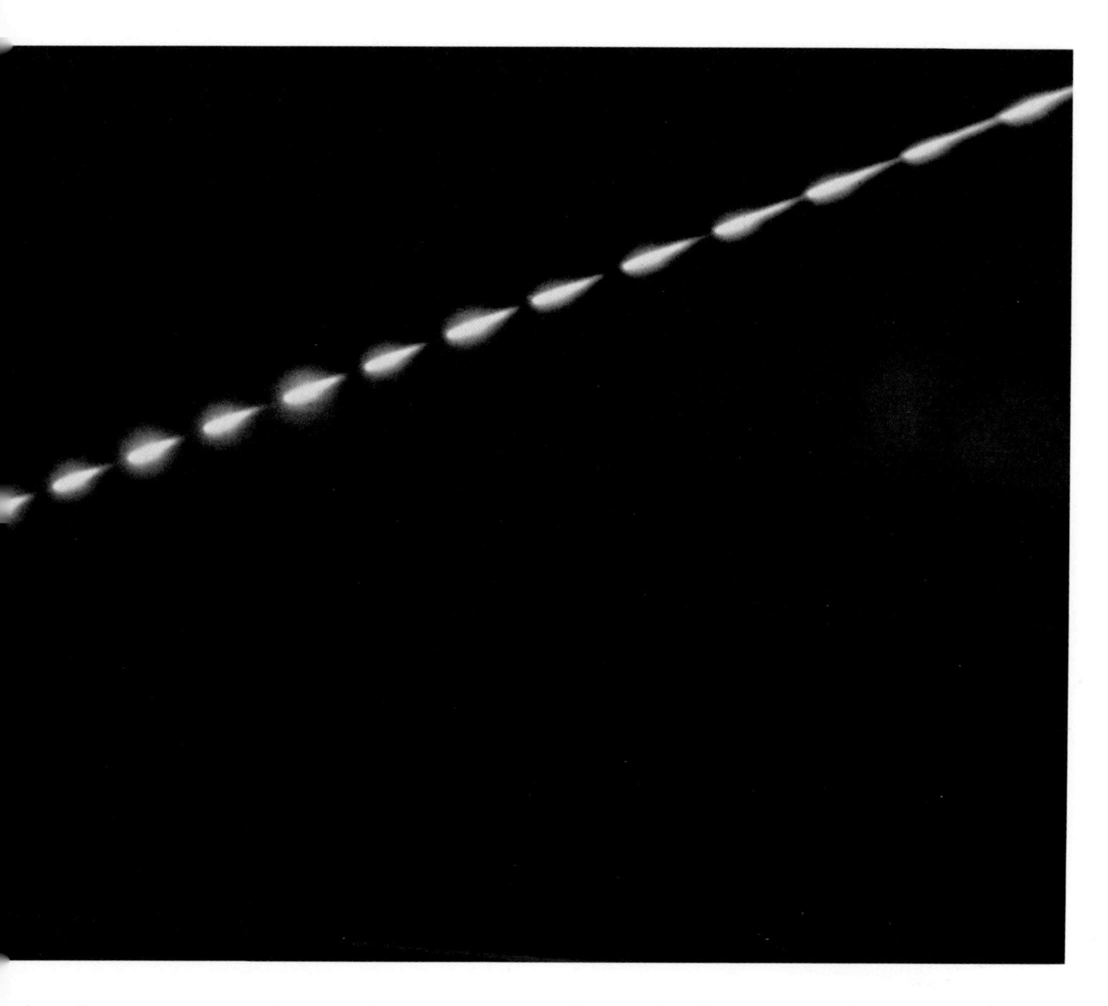

Fig. 98 *A MORP camera photograph of the fireball that presaged the arrival of the Innisfree meteorite (see Figure 61). The breaks in the trail were produced by a rotating shutter placed in front of the camera's lens, and allowed the meteoroid's velocity to be determined. (Image courtesy the Geological Survey of Canada and MIAC, http://miac.uqac.ca/MIAC/morp.htm)*

length can be made up. Since meteor photography is without doubt a very wasteful process, in the sense that the majority of images will not contain a single meteor trail, there is a definite cost saving to be made in buying bulk lengths of film. The other great advantage of using black and white as opposed to colour film is that black and white film is very easy to develop – and by doing this you will save yourself both time and money.

System Improvements

The initial velocities and atmospheric deceleration of meteoroids can be determined by photographic techniques. For these studies the camera system must be modified by mounting a rotating shutter in front of the camera lens. Such a rotating shutter system has the effect of breaking the meteor trail into a series of bright and dark dashes. Figure 98

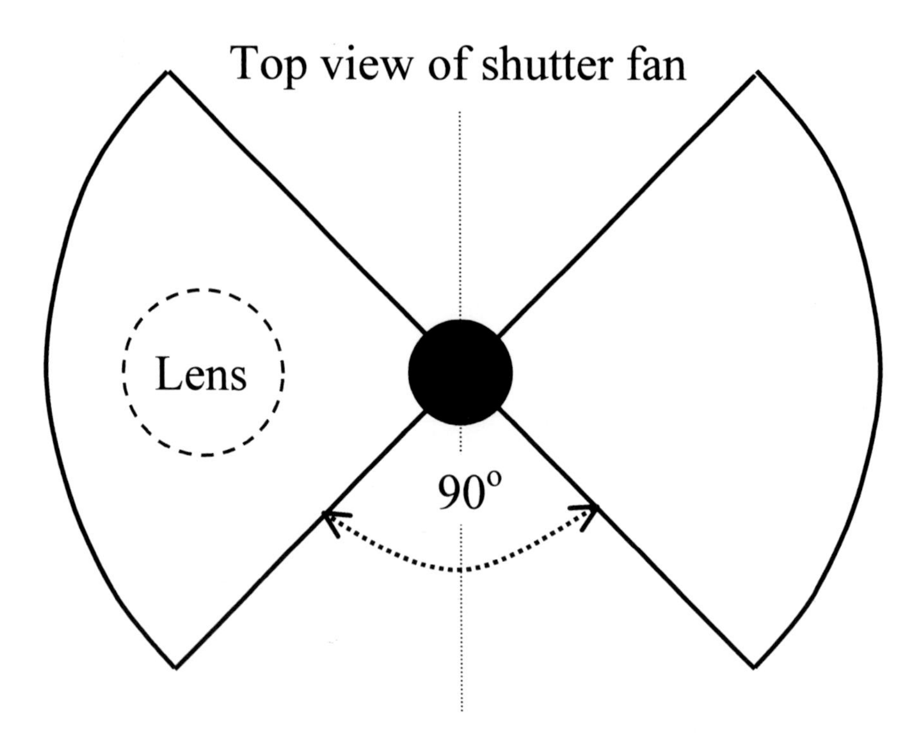

Fig. 99 *A typical rotating shutter system is made of a circular fan blade with two 90° cut-outs (as shown in the top view). For each rotation of the drive spindle, attached to the electric motor, the lens will be occulted twice. It is important that the shutter fan is well balanced and that the camera is isolated from any motor and drive-spindle vibration.*

shows a meteor trail that has been 'chopped' in this way by using a rotating shutter. The reason for breaking the trail image into a series of dashes is to determine the angular speed of the meteoroid as it descends through the atmosphere. This can be achieved by first determining the image sale of the photograph (from the recorded star positions and with the help of a star atlas), and then measuring the angular size L of each dash in degrees. The angular speed, ω, corresponding to each dash is then given by the ratio $\omega = L/T$, where T is length of each 'open phase' of the rotating shutter. The rotating shutter is usually driven at a speed that gives between 15 and 30 breaks per second (corresponding to dash duration times of between 1/30 and 1/15 of a second).

If you plan to build a rotating shutter system, then two important design features must be borne in mind. First you will need a motor that will run at a constant and known speed without excessive vibration. Second, the rotating fan must be well balanced and rotate without wobbling – remember that any large vibrations caused by a poorly mounted motor and/or an unbalanced fan blade will be transferred to the camera and will blur the images. One of the useful additional benefits of using a rotating shutter is that since the film is only exposed to light intermittently, longer exposure times can be used before fogging becomes an issue. Figure 99 shows a basic design for a rotating shutter.

To derive meteoroid speeds and decelerations from the angular speeds measured from the dash lengths and shutter duration, you will need to know the actual

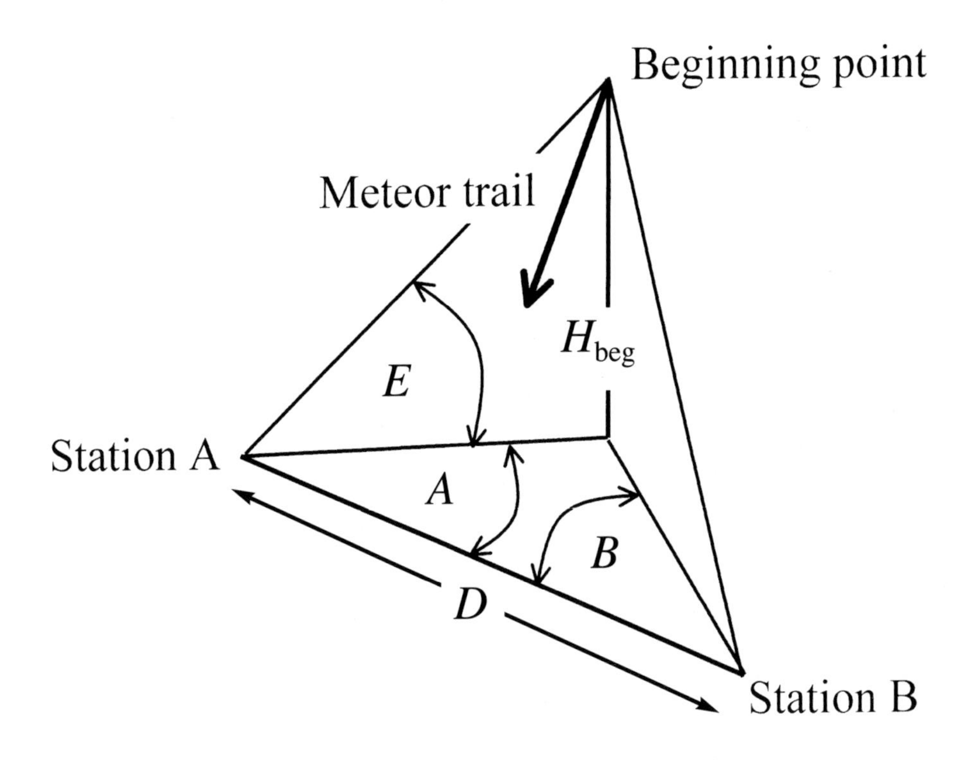

***Fig. 100** Determining a meteor's beginning height from two observing stations. The distance D between the two stations, A and B, must be known, and the elevation angle E along with the azimuthal angles A and B are determined from the photographic images and the known directions in which the cameras are pointing.*

orientation of the meteor's path through the atmosphere. As we shall now see, the meteor's atmospheric path can be determined by using two well-separated camera systems, of which only one need be equipped with a rotating shutter.

Beginning Heights

The basic principle behind the determination of meteor heights is to monitor a selected area of the sky with two camera systems separated by at least several tens of kilometres. When a meteor passes through the area being monitored, the beginning height can be determined by calculating two azimuthal angles and an elevation. Figure 100 illustrates the geometry for determining a meteor's beginning height.

The geometry is fairly straightforward to interpret. The beginning height, H_{beg}, is given by the expression

$$H_{beg} = \frac{D \sin B \tan E}{\sin [180 - (A + B)]} \quad (6)$$

where D is the known distance between the two observing stations. The end height, H_{end}, can be determined in exactly the same manner as the beginning height, using the elevation and azimuth positions of the meteor's end point in the calculation.

The actual analysis of the photographic images obtained from two stations is rather

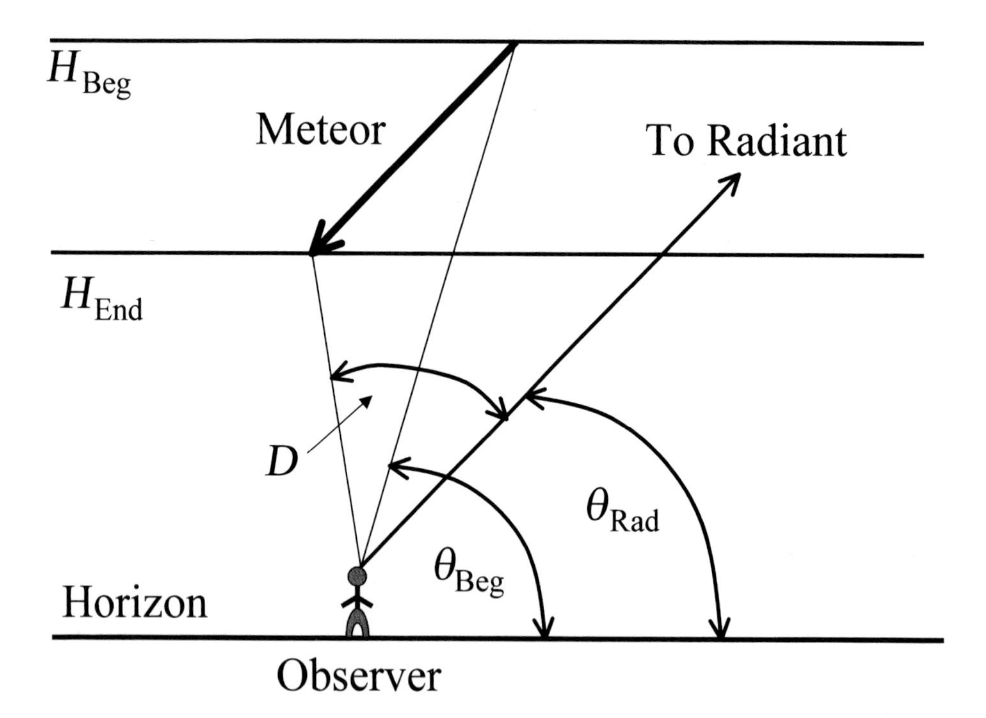

Fig. 101 *The relationship between the trajectory of a meteor in the atmosphere and the angles used to determine the beginning height.*

complicated, but the procedure can be made less time-consuming by using the AstroRecord computer program, developed by the International Meteor Organization and downloadable from http://www.imo.net/software/astrorecord/. By far the best procedure to follow if you are interested in determining meteor heights is to join forces with an established team of observers. There is always great merit is solving experimental and analysis problems for yourself, but there are also times when reinventing the wheel is unnecessary – get involved, ask questions and then develop your own techniques.

Observations of meteor trails with a one-camera system, either an ordinary camera equipped with a rotating shutter as described above or a single video camera, can yield approximate beginning heights provided one can be sure that a shower meteor has been imaged. This method works because the entry speed v is already known for shower meteoroids (*see* Table 2, p. 37). As long as a meteoroid did not slow down significantly during the luminous part of its atmospheric passage (which is true for small meteoroids), the angular speed ω can be estimated either from the broken image trail (still camera with rotating shutter) or by a frame-to-frame analysis of the video image. Here I shall describe the video analysis, but the photographic analysis procedure is exactly the same.

The angular extent of the trail, L, can be determined with the aid of a star atlas, and the number N of video frames in which the meteor is seen can be counted in still-frame playback mode. Each video frame is 1/25 of a second long (in Europe, which uses the Phase Alternating Line or PAL video standard), so the meteoroid's angular speed is given by ω =

L/(N/25). From the known time at which the meteor appeared (determined from the video clock), the location of the shower's radiant can be determined with the aid of a planetarium program. In addition, the elevation of the trail's beginning (θ_{Beg}) and end points are determined from the known direction in which the camera was pointing or with the aid of a planetarium program. If D is the angular distance (in degrees) on the sky between the radiant and the meteor trail's end point, then the atmospheric beginning height of the meteor, H_{Beg}, is calculated from the formula

$$H_{Beg} = \frac{57.296 \text{ v} \sin D \sin \theta_{Beg}}{\omega} \quad (7)$$

Provided the speed v is expressed in kilometres per second, equation (7) will give the beginning height H_{Beg} in kilometres. Figure 101 illustrates the geometry of the situation.

The expression for the angular speed given above is for video recordings captured at a frame rate of 25 frames per second, according to the European PAL format. In the National Television Systems Committee (NTSC) video standard used across North America the frame rate is 30 frames per second, and for recordings made in the NTSC format the angular speed must accordingly be evaluated as w = L/(N/30).

Meteor beginning heights depend upon the initial speed, mass and composition of the parent meteoroid. In this sense the beginning

Fig. 102 Beginning height plotted against meteoroid mass for four meteor showers. Fast meteors appear at systematically higher altitudes in the Earth's atmosphere than slower ones, and the beginning height typically increases with increasing mass. Geminid meteors appear to be the exception to this latter rule. (Based on P. Koten et al., Astronomy and Astrophysics, *vol. 428, p. 683, 2004)*

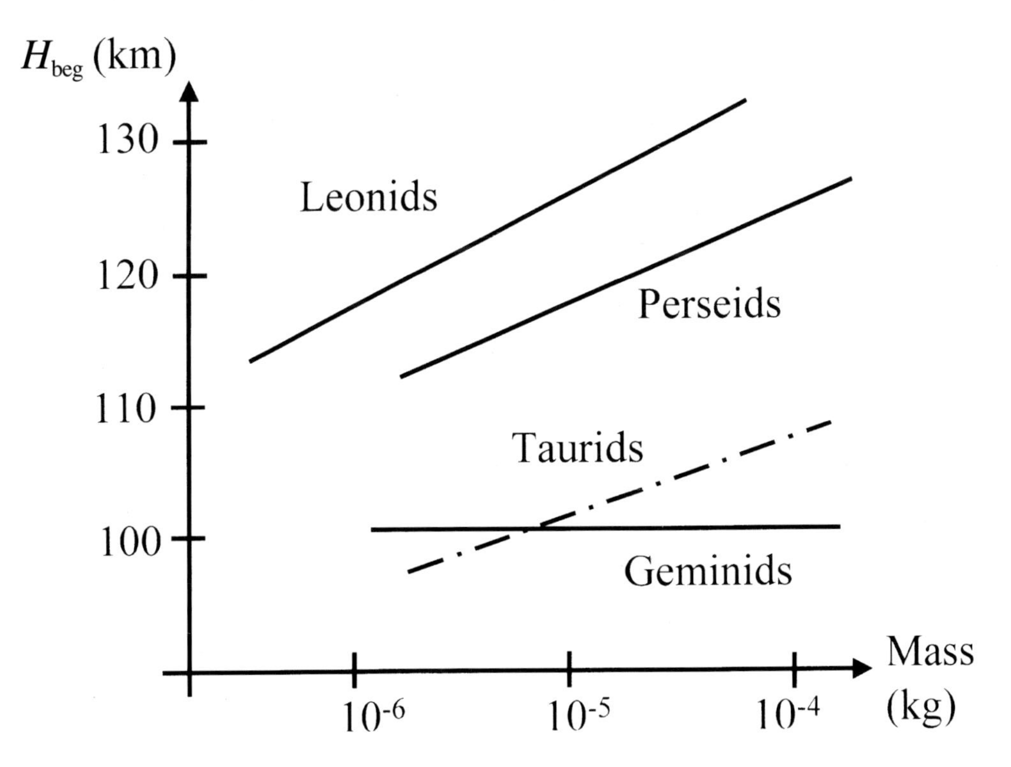

Fig. 103 *This 4-hour exposure taken with a fish-eye lens camera system on the night of the 1998 Leonid meteor shower's maximum captured 156 meteor trails. The faintest trail corresponds to a magnitude –2 meteor. (Image courtesy the Astronomical and Geophysical Observatory in Modra, Comenius University, Bratislava, Faculty of Mathematics, Physics and Informatics)*

heights of meteors from the same shower, which have the same initial speed and composition, should vary systematically with mass. One very useful project that could be contemplated with respect to the determination of meteor heights would be to look at the beginning heights for different meteor showers. During the recent Leonid meteor storms and outbursts, a number of observational groups reported the detection of extremely high beginning heights for some meteors – heights of 130km and more! This is an entirely new phenomenon, and it would be very interesting to see if the same holds for Perseid meteors, which enter the atmosphere at speeds similar to those of the Leonids. However, Pavel Koten (at Ondrejov Observatory in the Czech Republic) and his colleagues have found that the beginning heights of the relatively slow-moving Geminid meteors are essentially constant and independent of meteoroid mass. This result is no doubt related to the physical make-up of Geminid meteoroids and the fact that they are derived from an old cometary nucleus that has developed an inactive outer mantle.

Fish-Eye Lenses

The great advantage of using a fish-eye lens (*see* Figure 91, p. 100) is that the entire sky can be monitored with a single camera system. The disadvantages of fish-eye lenses are that they are quite expensive and they introduce a certain amount of distortion into the images. That said, a fish-eye lens is perfect for monitoring fireball activity. Figures 103 and 104 show two fine examples of fireball images obtained with fish-eye lens camera systems. All-sky images similar to Figure 104 (although most probably without a fireball present) can be obtained via the internet at the *Night Sky Live* network website, http://nightskylive.net/. The *Night Sky Live* network consists of eleven camera systems located at some of the world's most famous astronomical observatories, including the observatories at Hawaii, Cerro Pachon in Chile, Kitt Peak in Arizona, Mt Wilson in California, Siding Spring in Australia and on the Canary Islands. Each system uses an electronic CCD camera equipped with an 8mm fish-eye lens. The camera systems are fully automated and take a 180-second exposure of the sky at 4-minute intervals throughout the night.

Spectroscopy

The great importance of meteor spectroscopy is that it enables the composition of meteoroids to be determined. It is not the collection of the data, time-consuming though it is, that is the problem here, rather the complexity of the data analysis. Teaming up with a university-based researcher who can analyse any spectral images obtained is probably the best way to proceed.

As a meteoroid is heated and ablated during its descent through the Earth's

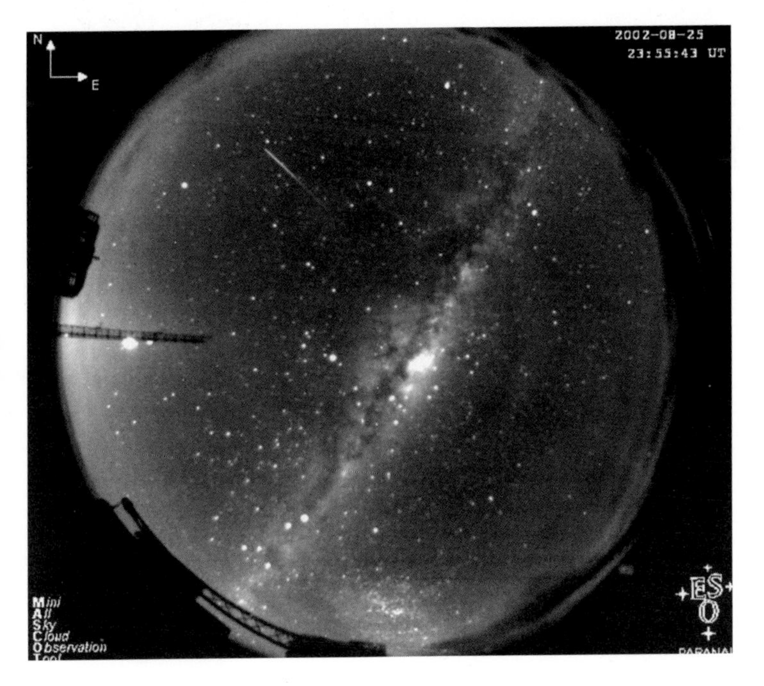

***Fig. 104** A bright fireball (upper left) was captured on 25 August 2002 in this image taken with the Mini All-Sky Cloud Observation Tool (MASCOT) operated at the European Southern Observatory at Parnal, Chile. MASCOT consists of a CCD camera attached to a fish-eye lens. The system automatically takes a 90-second exposure every 3 minutes, and the images are then used by astronomers to gauge the weather conditions. The Milky Way stretches from the lower left to the upper right in this image. (Image courtesy ESO)*

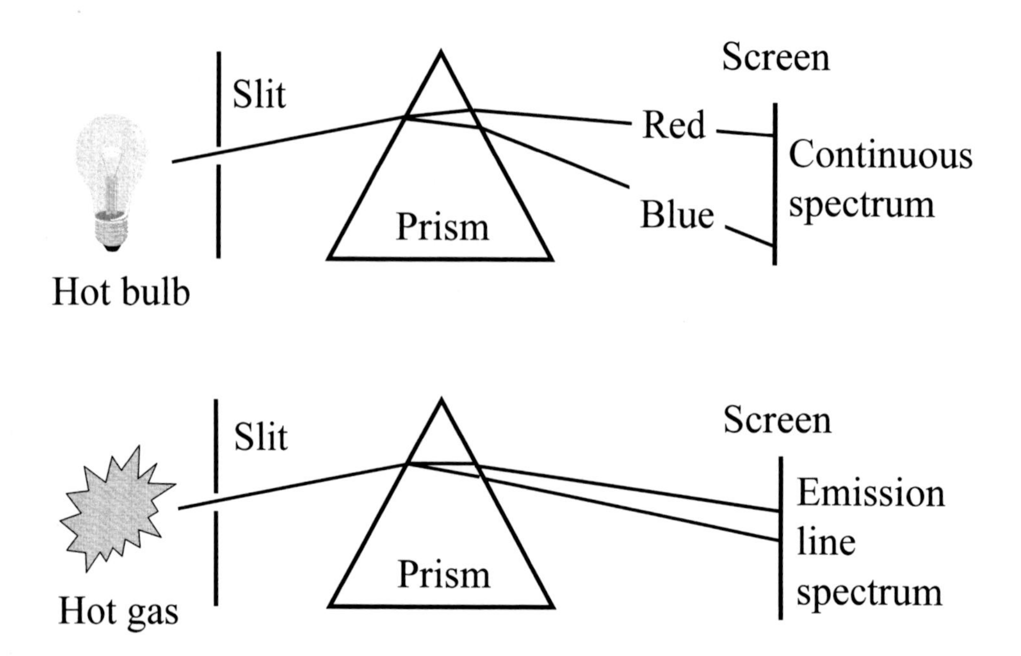

Fig. 105 A simple prism spectroscope. The triangular glass prism acts to spread the light that it receives from the entrance slit into its constituent colours. Light from a hot lamp produces a rainbow of colours (i.e. a continuous spectrum) – the light is spread out over all possible visible wavelengths from red to blue. Light from a hot gas, however, produces only distinct emission lines in just a few well-defined colour (or wavelength) regions.

atmosphere, its constituent atoms become ejected from its surface and eventually suffer collisions with atmospheric molecules. This results in the production of an **emission line spectrum**. The key characteristic of an emission line spectrum is that the light is produced at very specific wavelengths or colours. This is in contrast to the spectrum produced by an ordinary light bulb, which is a continuous spectrum. Viewed through a spectroscope, a light bulb will produce the familiar rainbow pattern of light, but a hot gas (such as that surrounding a meteor) when viewed through the same spectroscope will produce a series of individual bright lines on a dark background. Figure 105 illustrates the essential difference between a continuous and an emission line spectrum.

That meteors produce emission line spectra indicates that the light we detect from them must be derived from a hot gas, and not from an incandescent solid body (although the atoms and molecules that make up the meteor 'gas' do of course come from the meteoroid itself). Key to the interpretation of emission line spectra is that each type of atom and molecule produces its own unique set of emission lines. The emission lines from, say, magnesium can be unambiguously distinguished from those of, say, calcium. Spectroscopy is primarily concerned with the identification of those atoms responsible for the emission lines that are present in a given spectrum. In short, it is like fingerprinting – only one set of suspects can produce the lines that are observed. Figure 106 shows the spectrum of a magnitude –5

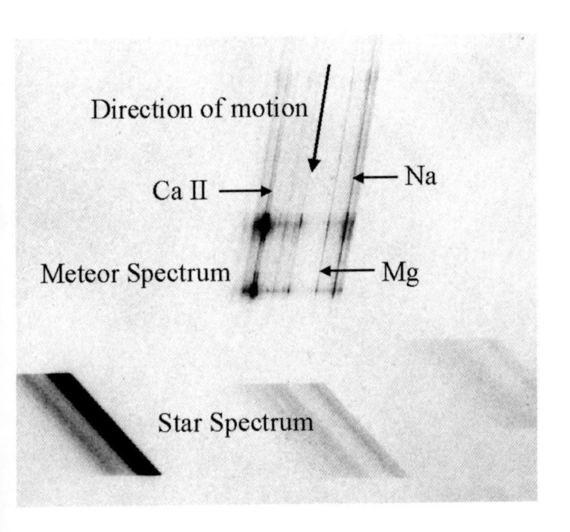

Fig. 106 *The emission line spectrum of a Perseid meteor captured on 13 August 1989. The image was obtained with a modified large-format F-24 aero camera with an f/2.8, 200mm focal length lens. The spectrum was recorded on a 4 × 5-inch sheet of Kodak Royal Pan 4141 film. A detailed analysis of the spectrum was published by Borovicka and Majden in the* Journal of the Royal Astronomical Society of Canada, *vol. 92, p. 153, 1998. (Image courtesy E. Majden)*

Perseid fireball captured by Canadian amateur astronomer Ed Majden. The spectrum was obtained with a prism spectrograph, and an analysis by Jiri Borovicka at the Ondrejov Observatory revealed that emission lines due to magnesium, calcium, iron and sodium were present in the spectrum.

Meteor spectra can be obtained with a basic camera system and an appropriate prism or transmission diffraction grating. A diffraction grating consists of a very finely ruled flat piece of glass (or clear plastic) with thousands of fine parallel grooves, or lines, etched into its surface per centimetre. The effect of the parallel grooves is to spread the incident light into its constituent colours, just as a prism does. So-called blazed diffraction gratings are preferable to prisms for meteor work because they spread the meteor's light into a brighter and more uniform spectrum than a prism can (which helps in analysing the spectral lines) and, because they are quite thin, they can capture spectra of fainter meteors. This latter characteristic is certainly useful because there are more faint meteors than bright ones. The one big drawback with using

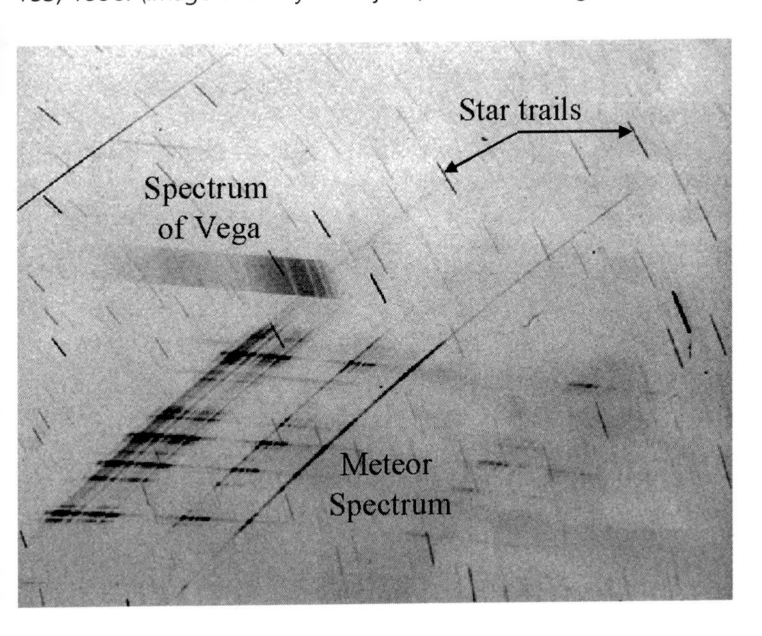

Fig. 107 *The spectrum of a magnitude –4 fireball captured on 9 June 1997 with a holographic grating. The image was obtained with a Bronica large format camera with an f/2.8, 75mm focal length lens (set to an f-stop number of 4.0), using Kodak Tri-X Pan 120 roll film. The spectrum of the bright star Vega is also indicated. (Image courtesy E. Majden)*

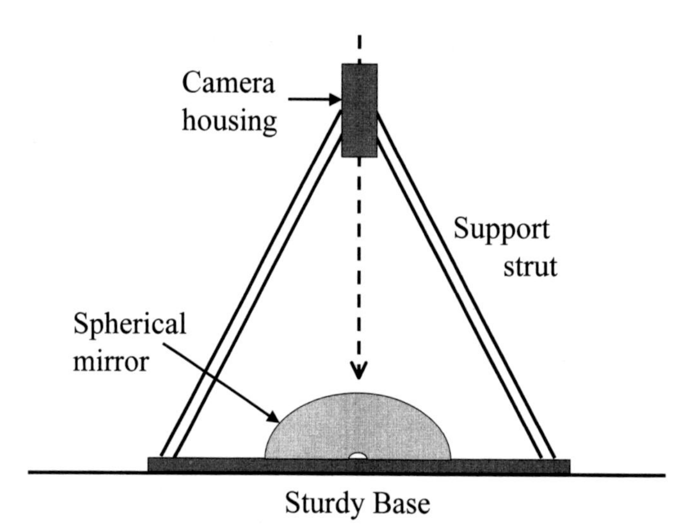

Fig. 108 *A basic all-sky video system design. The video camera is placed in a watertight housing and points directly down towards the centre of the curved mirror. The zenith is obviously obscured by the camera itself, but the rest of the sky is visible to the camera. Four struts are used to hold the camera housing steady and in place. The physical dimensions of the system will depend upon the diameter of the mirror used and the camera lens employed.*

diffraction gratings, however, is their cost – a good grating is not cheap. An inexpensive alternative to using a grating is to purchase a section of a holographic grating film. This film can be obtained from any educational supplier and is certainly affordable. The holographic grating film can be bought in sheets, or more usefully as ready-mounted 35mm slides. Figure 107 shows another spectrum captured by Ed Majden, this one with the aid of a holographic grating.

The procedure for meteor spectroscopy is quite straightforward. The camera and grating system is aimed at a chosen sky location, and a series of long-exposure images secured on as many clear nights as possible. After that, be patient. While the chances of successfully capturing a good meteor spectrum are certainly small they are not zero, and sooner or later an image will be captured. Milos Weber, a long-time meteor observer in the Czech Republic, reports, for example, that in a cumulative survey time of 1,224 hours (over 202 nights between 1994 and 2003) a total of six meteor spectra were captured – that's 204 hours of observing time per spectrum.

Since the diffraction grating will severely limit the sensitivity of the camera system (i.e. only very bright meteors will produce a measurable spectrum), use a fast film and the longest exposure time you can – indeed, push the exposure time to the limit where fogging just begins to occur. Some trial-and-error experimentation at your specific observing location will be required to determine the longest achievable exposure time.

A full theoretical understanding of meteor spectra is still waiting to be formulated. It is clear, however, that the emission lines that are observed are derived from a hot gas, at a temperature of about 4,000°C, that surrounds the ablating meteoroid. Emission lines from many diverse atoms and molecules have been observed in meteor spectra. In most spectra the strongest lines are due to the so-called H and K lines of ionized calcium, and lines due to sodium and magnesium. Spectra obtained from very-high-speed meteoroids (such as those that produce Leonid meteors) also show emission lines due to atmospheric oxygen atoms and nitrogen molecules. The relative abundances of magnesium, calcium, iron and sodium in Leonid meteoroids have recently been found to be similar to those derived for carbonaceous chondrite meteorites. Some meteors, however, show spectra that are entirely deficient in sodium, and the reasons for this are currently a mystery. There is still a great

Fig. 109 The all-sky video camera system that I operate at Campion College in Regina, Saskatchewan, Canada. Meteors of magnitude –3 or brighter are routinely captured on video with this system. The camera forms part of the Sandia Fireball Network of cameras distributed across the North American continent.

deal of work that needs to be done in the area of meteor spectroscopy, and any new spectra that might be captured will be of great scientific value.

An All-Sky Fireball System

An efficient way of monitoring fireball activity is to use a video system with the camera mounted above the centre of a curved mirror. Figure 108 illustrates a basic design. Such a camera system will provide both the time of occurrence, the time of flight and positional information on any fireball that crosses the sky.

The specifications for the all-sky camera system I use in Regina (*see* Figure 109) are as follows. The hemispherical mirror has a diameter of 46cm, and the four support struts, of length 118cm, have their lower ends attached to two cross-braces of length 92cm. The camera housing is an 11cm diameter section of PVC pipe. The camera is located about 85cm above the highest point of the mirror. This height will vary according to the type of mirror and lens that you use. The key is to experiment with the equipment and adjust the camera height until the base of the mirror fills most of the field of view. The camera can be any standard video camera rated for outdoor use – my system employs a Computar high-sensitivity monochrome (model FC-08B) camera with an 8mm, f/1.2 lens coupled with an automatic iris adjustment system. The mirror need not have any special finish, and this one is simply an off-the-shelf security mirror as used in many retail shops. The most important feature of the mirror is that, when viewed from above, its curved surface should provide a complete view of the sky.

The most important feature of any all-sky camera system is its ease of operation. The system must be sturdy, capable of continuous operation in all weathers and simple to operate. A system that requires continual repair and maintenance will soon be a system that is abandoned and unused.

The video output from the camera system is recorded onto standard VHS videotape, and it is convenient to use two video recorders so that as much of a night-time session as possible can be recorded. Use the VCRs' automatic programming feature to stagger the start times for each tape. Tapes need be examined only if a fireball report is made, or alternatively one can use a meteor recognition program, such as *MetRec* ('Meteor Recognizer') developed by Sirko Molau, to log the times of events. The *MetRec* program can be used in real time or for off-line inspection of previously recorded videotapes. The software for *MetRec* can be downloaded as a shareware program from http://www.metrec.org/. The system requirements call for a PC with a Pentium processor and a Matrox Meteor II frame grabber (http://www.matrox.com/imaging/products/frame_grabbers.cfm). The software and frame grabber can accommodate both PAL and NTSC video standards and runs in DOS mode (it does not run under the Windows environment or on Macintosh computers). Once configured, *MetRec* will automatically record the times at which any fireballs appear.

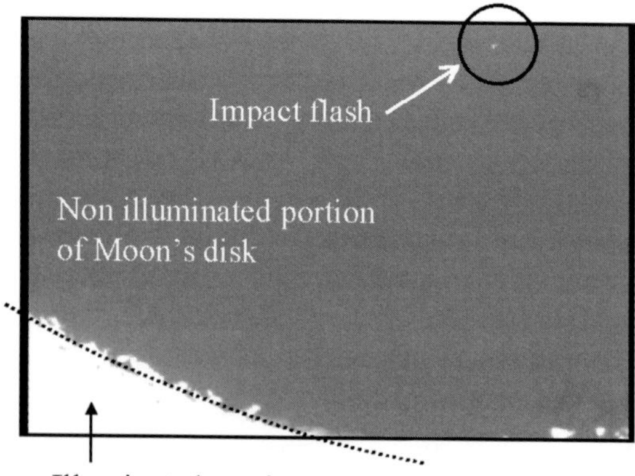

Fig. 110 *A single frame from a video sequence showing the impact of a Leonid meteoroid on the Moon, captured on 19 November 2001 at 00:18:58 UT. The impact produced a flash of light lasting about 1/10 of a second. The Moon's illuminated portion is visible as the curved line at the lower left of the frame. The observations were made with a 125mm aperture Celestron C5 telescope, and the video data were analysed with a program called VideoScript (www.videoscript.com).*
(Image courtesy David Palmer)

Serendipitous Video

We live in an era in which video cameras and cellphones that can send pictures are common items. This popular obsession with gadgetry, however, has produced some useful scientific payback, and images of many serendipitously observed fireball events are now being obtained. Fireballs are often so bright that even the most basic of video camera systems can record their flight under both night- and daytime conditions.

If you chance to witness a fireball and you have a video camera at the ready, then try to record a steady image – if possible, lean against a wall or a tree to steady the camera. Try to capture some distant objects (trees, mountains or buildings) along with the fireball in the field of view since this will enable the fireball's direction of motion to be determined from the video images. Make sure that you clearly record where you were standing when the video sequence was taken, since you will need to use the exact location as a reference point for determining the bearings of any distinct landmarks captured on video. Do not zoom in on the fireball: a zoomed-in image will exaggerate any camera shake, and there will be less chance of foreground reference features appearing in the image.

The fall of the Peekskill meteorite (*see* Figure 49, p. 49) on 9 October 1992 is a classic example of what can be achieved if enough video frames can be collected. The fireball associated with this meteorite fall was seen to race across the skies of the North-eastern United States in the early evening hours on a 'Friday football' night (*see* Figure 58, p. 57). Many serendipitous video records of the fireball were captured by observers watching and video recording their local games. A team of researchers eventually tracked down sixteen videos of the fireball, and the atmospheric path and orbit of the meteorite were determined

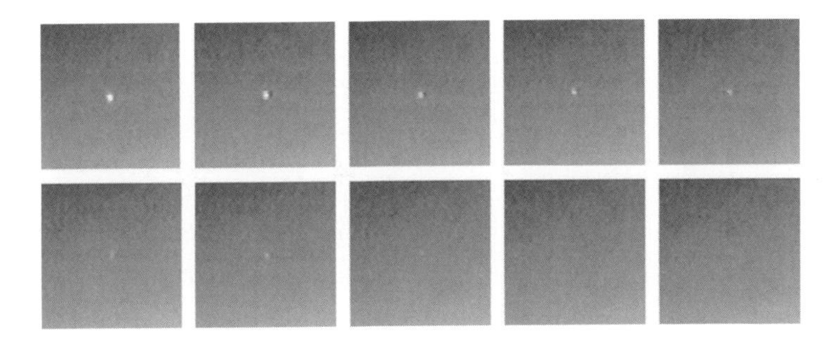

Fig. 111 A video sequence showing the flash produced by the impact of a Leonid on the Moon's surface. The frames are 1/60 of a second apart. The observations were gathered with a CCD video camera attached to a 200mm diameter Newtonian reflector. The frames were transferred to a Hi-8 video recorder and later analysed with a program that compared brightness changes in successive video frames. (Image courtesy Masahisa Yanagisawa)

from the data. A selection of Peekskill videos can be viewed at the following website: http://aquarid.physics.uwo.ca/~pbrown/Videos /peekskill.htm/.

Shooting the Moon – Lunar Impacts

Just as the Earth's atmosphere is continuously struck by meteoroids, so too is the surface of the Moon. While the Moon has no atmosphere in which to produce meteors, the surface impacts themselves can generate light. For many years lunar observers have reported seeing impact flashes, but the first video observation of an impact flash was recorded during the 1999 Leonid storm. Figure 110 shows a single video frame from a later sequence of meteoroid impacts on the Moon.

Japanese observers Masahisa Yanagisawa and Narumi Kisaichi of the University of Electro-Communications, Tokyo, recorded five Leonid impacts on the Moon on the night of 18 November 1999. Figure 111 shows a dramatic sequence of images relating to one of these events, which produced a magnitude +3 flash at 13:54:26.0 UT. The deduced time development of the impact flash is shown in Figure 112, where it can be seen that the light curve has a very rapid rise to a maximum and then a relatively slow decay. It has been estimated that the flash was produced by a Leonid meteoroid of about 3kg crashing into the Moon's surface. The event probably produced an impact crater about 10m across.

Monitoring the Moon's surface for meteoroid impact flashes can be done only when the Moon is partially illuminated as seen from Earth, as shown in Figure 113. Also, since the impact flashes will be brighter for higher impact speeds, the best times at which to conduct lunar impact surveys will be during showers such as the Perseids, Orionids and Leonids, which all have impact speeds of 60km/s or more. The Association of Lunar and Planetary Observers (ALPO) maintains a very useful web page that provides information on the best times to look for lunar impacts (http://www.lpl.arizona.edu/~rhill/alpo/lunar stuff/lunimpacts.html). Only a very few impact flashes from non-Leonid meteoroids have ever been recorded, so there is a clear opening for the keen amateur to make significant contributions.

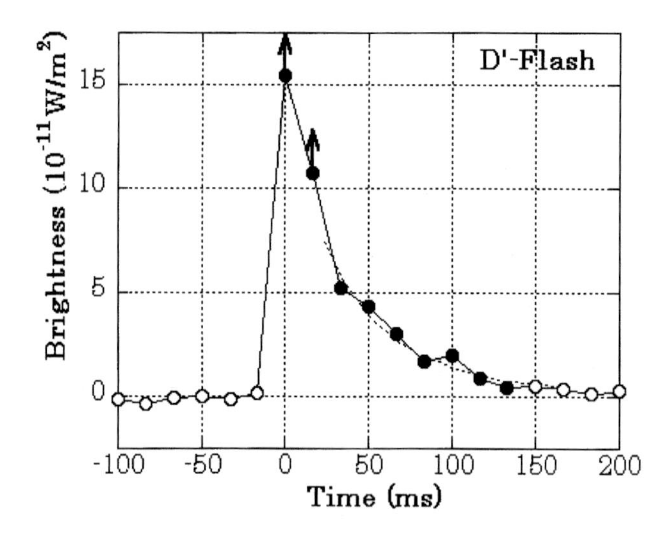

***Fig. 112** The time development of the impact flash (designated D') shown in Figure 111. The data points (shown as filled circles) are 1/60 of a second apart, and the y-axis shows the measured energy flux (brightness). The arrows indicate that some of the pixels of the CCD became saturated, and consequently the energy flux is an underestimated value. The open circles are background calibration points from video frames just before and just after the impact. (Image courtesy Masahisa Yanagisawa)*

The video system I use to look for lunar flashes is shown in Figure 114. The system consists of an Astrovid 505E black and white CCD video camera attached to a 150mm Celestron telescope. The system tracks the sky to compensate for the Moon's motion, and the video output is recorded onto standard VHS videotape. The observing time is recorded onto the video image with a video time/date clock. A standard video camera with a zoom feature can also be used to monitor the Moon for impact flashes if it is on a sturdy tripod mount. In this case, simply zoom in on the Moon's disk as far as possible and track the Moon's motion as best you can.

Perhaps the biggest problem in detecting impact flashes is their very short duration – one blink and you may have missed it. I usually monitor the video image as it is being recorded, and also review the tape again at a later date. Fatigue soon sets in during this monitoring procedure, so work in short sessions of 15 to

***Fig. 113** The optimum geometry of the Earth, Moon and radiant for detecting meteoroid impact flashes upon the Moon's surface (not to scale). The lunar phase must be such that the terrestrial observer can see part of the non-illuminated lunar hemisphere, and the radiant must be in a direction well away from the direction to the Sun. The smaller arrows in the diagram represent meteoroids that could produce light flashes visible to a terrestrial observer. (Based upon a diagram published in the journal* Il Nuovo Cimento, *vol. 21C (5), p. 577, 1998)*

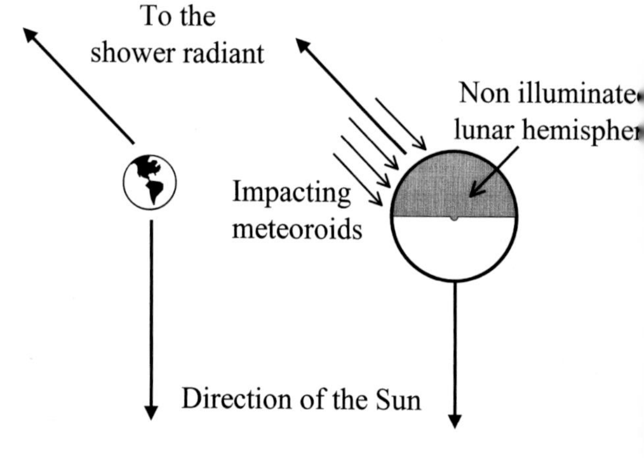

20 minutes, and take regular 10-minute breaks. There is little point in monitoring the live image or tape replay if you are tired or distracted. Ideally, a specialized computer code could be developed to perform the monitoring task, but this would certainly require specialist software design skills, and to my knowledge no such software is commercially available.

Fig. 114 *A video camera attached to a telescope system for monitoring the Moon's surface for impact flashes. The telescope is attached to a motorized equatorial mount, and the camera is attached to the telescope in such a way that the camera's CCD detector is at the telescope's prime focus.*

Flickering

Visual observers occasionally report a flickering phenomenon in which the trail of a fireball shows rapid small-scale oscillations with time. The variation is typically just a fraction of a magnitude. This effect is real, not illusory, and has been recorded on both photographic images and video sequences. Figure 115 illustrates this phenomenon as measured for a magnitude –9 Geminid fireball recorded on 13 December 2002 with the all-sky video camera system at Regina in Saskatchewan. For this particular fireball the flickering spanned about a tenth of a magnitude, each oscillation lasting about one-sixth of a second.

In a detailed study of fireball data I carried out with Peter Brown at the University of

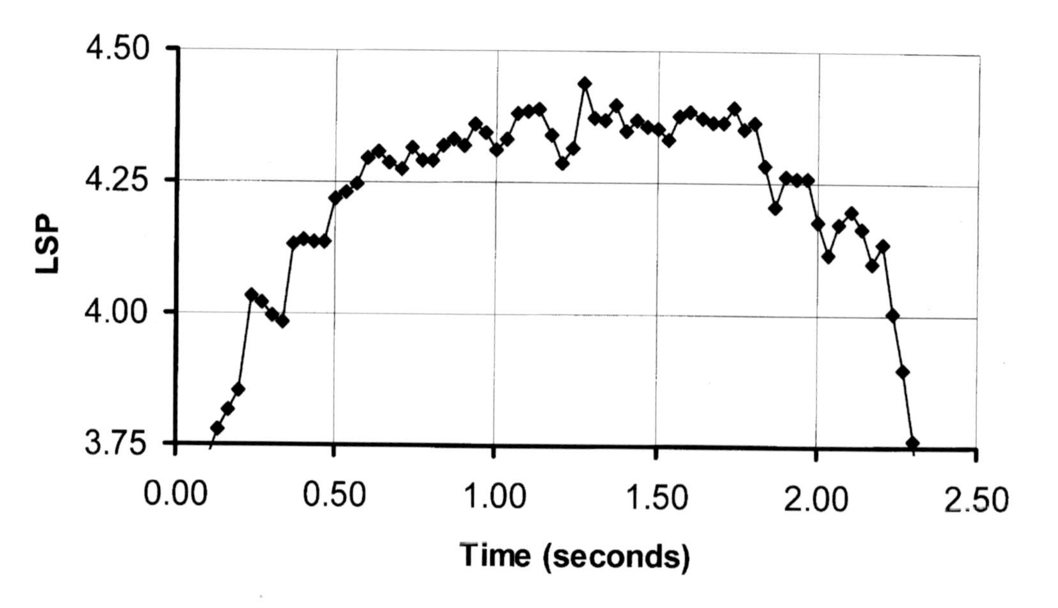

Fig. 115 *A light curve of the flickering variations in a magnitude –9 Geminid fireball seen on 13 December 2002. The curve shows the flickering as derived from a video sequence sampled at a rate of 30 frames per second. The y-axis is the brightness expressed in terms of the measured log sum pixel count (see p. 125).*

Western Ontario, we found that something like 5 per cent of the fireball trails in the MORP survey showed a distinct flickering. Also, about 70 per cent of Geminid meteors brighter than magnitude –3 that I have detected with the all-sky camera system in Regina show a flickering effect. In contrast, for the same magnitude limit, 18 per cent of sporadic fireballs, 0 per cent of Perseid meteors, and 4 per cent of Leonid meteors show a flickering effect. Clearly there is something about Geminid meteoroids that makes them prone to flickering, while the Perseids, in contrast, never appear to flicker. This difference is probably related to the initial speeds of Geminid and Perseid meteoroids when they enter the atmosphere (35 and 65km/s, respectively), and to distinct differences in the physical constitution of the meteoroids. The high initial speeds of Perseid meteoroids, for example, result in the 'smearing-out' of any brightness oscillations. So the zero percentage of Perseid fireballs showing flickering is most likely to be an observational selection effect rather than a real physical phenomenon. Likewise, the orbits of Geminid meteoroids periodically take them close to the Sun, and this, it has been suggested, causes them to become 'baked and hardened' – a condition that allows them to be spun up (by a 'windmill' effect driven by solar radiation) without them bursting. Perseid meteoroids, in contrast, are believed to be more fragile than Geminid meteoroids and consequently are more difficult to spin up without them being disrupted.

Video observations are ideal for studying flickering since the individual video frames give a time-lapse sequence of events. The frame-by-frame analysis of a video sequence is a straightforward procedure provided one has access to a good image analysis program. One excellent program that I have used is *NIH*

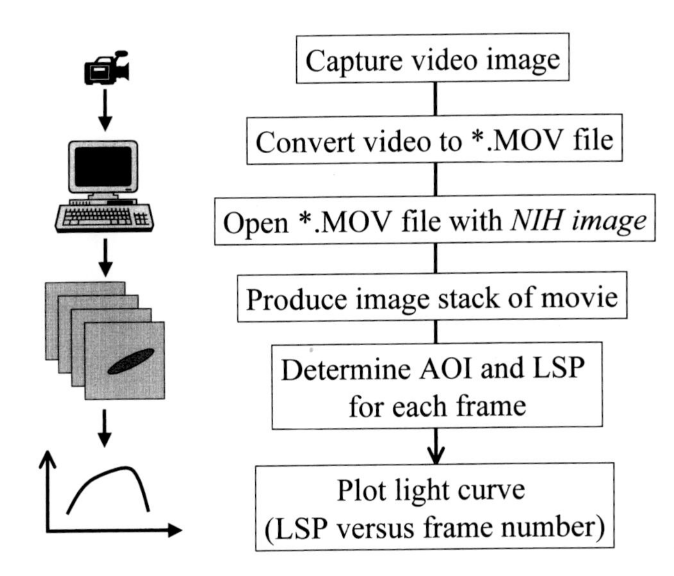

Fig. 116 *The various steps in determining a time sequence of log sum pixel (LSP) values. Once an LSP value has been determined for each video frame, a light curve of the meteor trail (plotted as LSP versus frame number) can be constructed, as in Figure 115.*

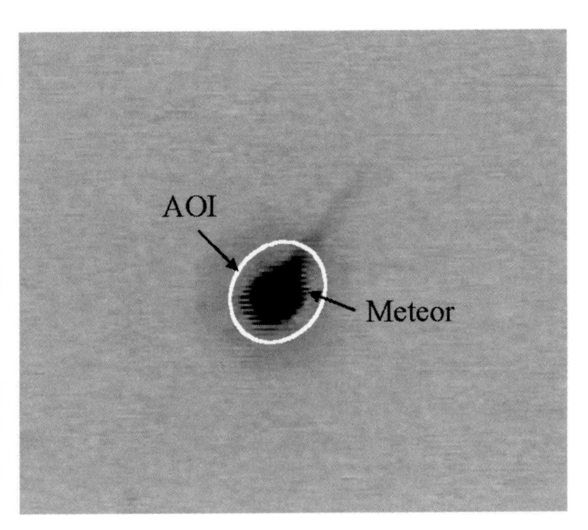

Fig. 117 *A single-frame image of a meteor and its associated area of interest (AOI). When tracing out the AOI, always try to avoid any background stars, since these will add to the pixel sum and artificially increase the brightness of the meteor.*

Image. This particular program was developed for the National Institute of Health in the United States and is available as a freeware download from http://rsb.info.nih.gov/nih-image/. Download and install the latest version of the program and then experiment with its many features. *NIH Image* runs only on Macintosh computers, but for Windows and Linux there is another NIH program, called *ImageJ*. There are many commercially available programs that one can purchase and use to perform image analysis, but few of them are actually as good as *NIH Image*.

The analysis of a video sequence proceeds as shown in Figure 116. Open the *.mov file of the fireball sequence to be studied with *NIH Image* and then convert the movie sequence into an image stack (a series of individual frames superimposed upon one another); there is a specific *NIH Image* command for this procedure. Then, analyse each image in the stack in turn. First, systematically trace around the meteor to produce an area of interest (AOI), and then determine the pixel sum (PS) for the AOI – this information is automatically evaluated by *NIH Image*. The PS is simply the sum of all the pixel values that fall within the boundary of the AOI (*see* Figure 117). Various studies have shown that a meteor's magnitude is actually proportional to the log sum pixel count (LSP), which is the logarithm to base 10 of the pixel sum in the AOI.

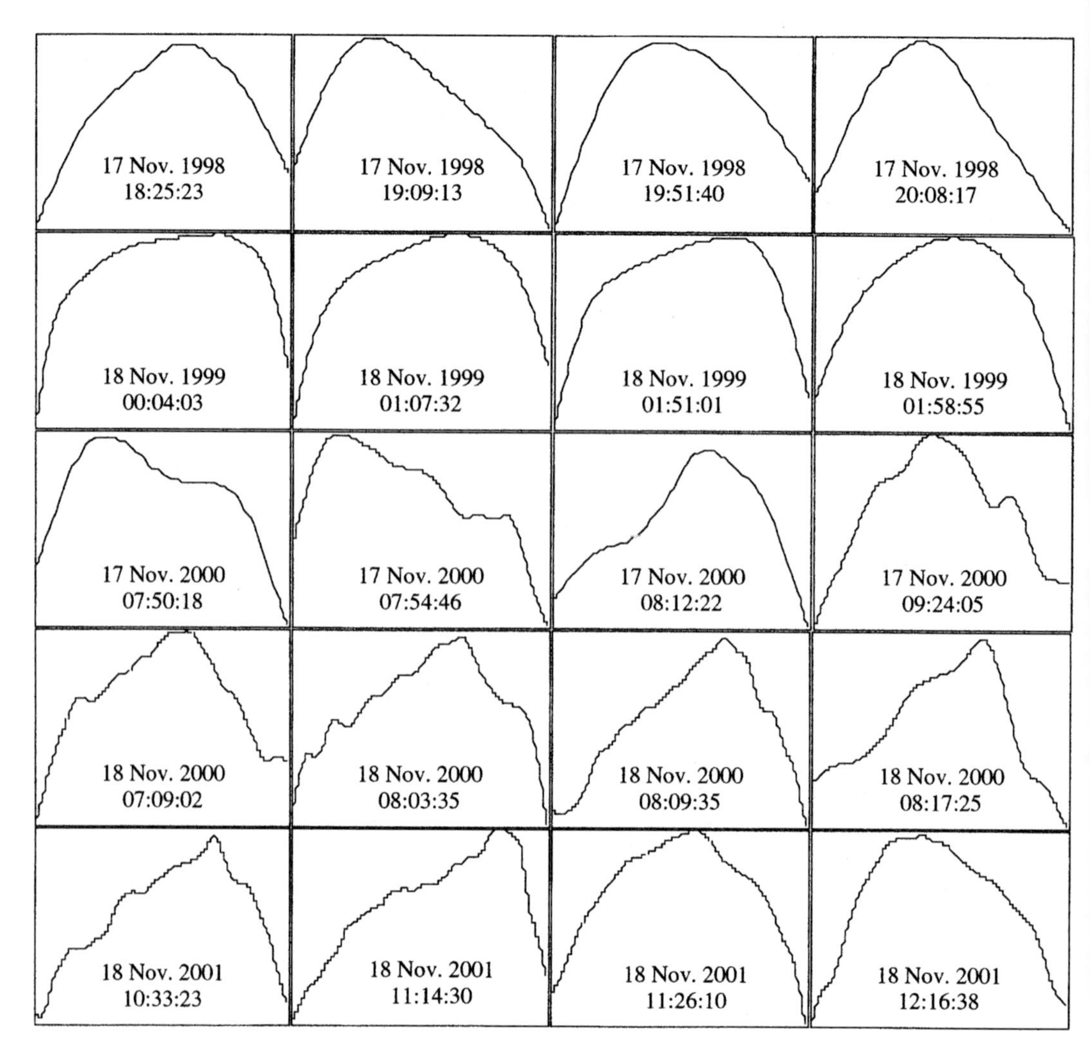

Fig. 118 *A selection of Leonid meteor light curves. To make comparison easier, the curves have been scaled so that the brightness and duration of each of them vary between zero and one. Although the meteoroids are all from the same comet (comet Tempel–Tuttle), the observed meteor light curves can be early peaked, symmetrical or late peaked. Some light curves also show secondary peaks and linear segments. (Image courtesy Ian Murray, University of Regina, based on Beech and Murray,* Monthly Notices of the Royal Astronomical Society, *vol. 345, p. 696, 2003)*

Once the LSP has been obtained for each frame in the image stack, then a **light curve** can be constructed. The light curve is a graph showing how the meteor's brightness varies with time. Since the time interval between each video frame is 1/30 of a second (NTSC standard; 1/25 of a second in the PAL system) the light curve can be constructed by plotting the LSP value against the image frame number. There has been much recent interest in the study of meteor light curves, since the moment of maximum brightness along the trail provides clues to the physical structure of the meteoroid.

A solid, monolithic grain will always produce a light curve that is brighter towards the end of the trail – a so-called late-peaked curve. A meteoroid made of many small grains will produce a symmetric or early-peaked light curve, depending upon the size distribution of the grains.

Figure 118 shows a selection of meteor light curves recorded by an image intensified video camera system (*see below*) during the Leonid meteor showers in each of the years from 1998 to 2001. There is a great variation in the shapes of these curves, and this is clear evidence that Leonid meteoroids are not single, solid grains, but must each be made of many smaller particles of differing mass – conforming to the 'dust-ball' meteoroid model (*see* Figure 3, p. 10). The shape of the light curve varies according to the specific number and mass range of the particles that make up the meteoroid. The systematic study of light curves from the various annual meteor showers would make an excellent research project.

Image-Intensified Systems

The limiting magnitude accessible to a standard video camera is not particularly impressive – only meteors of perhaps zero magnitude and above can be detected. This limit can be significantly improved upon, however, by coupling the camera to an image intensifier. The image intensifier literally acts as a light amplifier, boosting the light output to the camera system considerably. A typical second-hand, army-surplus image intensifier coupled

Fig. 119 *The basic layout of an image-intensified system. The objective lens gathers the light from the meteor and brings it to a focus on the entrance window of the intensifier. This produces a 'shower' of electrons, and these electrons are multiplied and accelerated towards the exit fluorescent screen. When the electrons strike the fluorescent screen, an image is formed that is many times brighter than the optical image that would be produced by the objective lens alone. A coupling lens is used to bring the image formed on the fluorescent screen to a focus on the CCD chip in the video camera.*

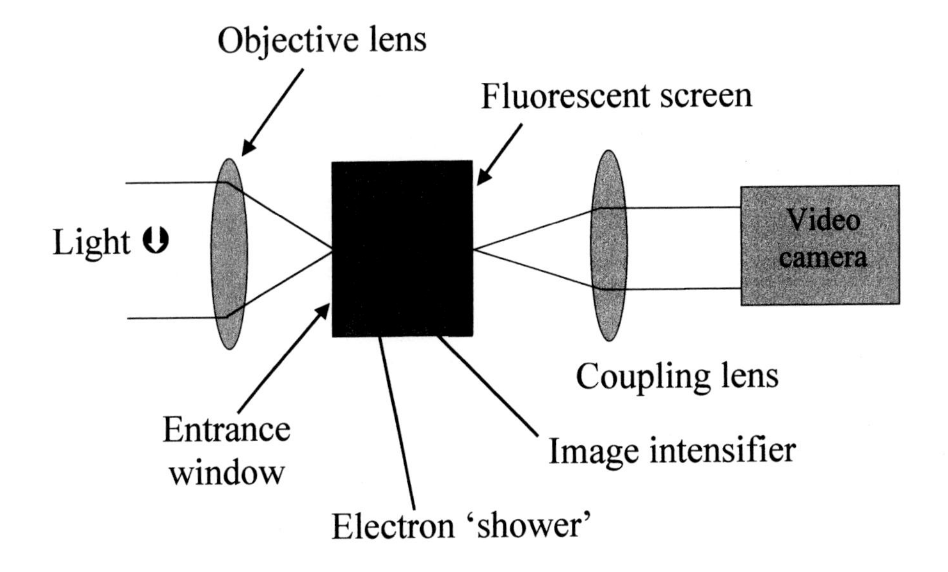

to a video camera will probably enable meteors as faint as magnitude +5 to be detected, and newer intensifiers (so-called Gen III systems) can push this limit closer to magnitude +7. There is a price to pay for all this increased light-gathering sensitivity, however: while the time resolution of the video observations remains good, the image resolution becomes decidedly poor – indeed, the resolution is much poorer than what is routinely achievable by standard photographic techniques. The great advantage of using an image-intensified system is that far more meteors are likely to be detected – there are many more faint meteors than bright ones. In recent years image-intensified systems have been used to great effect in studying the light curves of faint meteors and in meteor spectroscopy.

7 Radio Observations

Optical observations of meteors, whether with the naked eye or with cameras or other detectors, are subject to a number of natural and unavoidable selection effects. Clearly, when the Sun is above the horizon it is not worth pursuing visual meteor counts. Likewise, for about ten nights in each month it is hardly worth conducting visual observations because the Moon is too bright, and of the remaining nights perhaps only one-third will be clear. In short, there is much to frustrate the efforts of even the keenest visual meteor observer. All these frustrations can be circumvented, however, by turning to radio techniques. Through streaming sunlight, pouring rain, clouded skies and a full Moon, the radio eye can 'see' meteors. Indeed, meteor astronomers have been making radio and radar observations of meteors and meteor showers for the past sixty years, building directly upon the development of the technology since the end of World War II. This chapter covers the basic requirements for the construction of a '24/7' radio-meteor monitoring system.

Physics Background

James Clerk Maxwell was the first to demonstrate that light is a form of electromagnetic radiation, in 1865. He showed that what our eyes perceive as light is really the propagation (at the speed of light) of a wave that has an electric and a magnetic component. Building upon the theory developed by Maxwell, Heinrich Hertz designed an experiment that demonstrated the existence of what we now recognize as radio waves, which are another form of electromagnetic radiation. Indeed, the familiar X-ray, ultraviolet (UV), infrared (IR) and microwave forms of radiation are all electromagnetic waves that

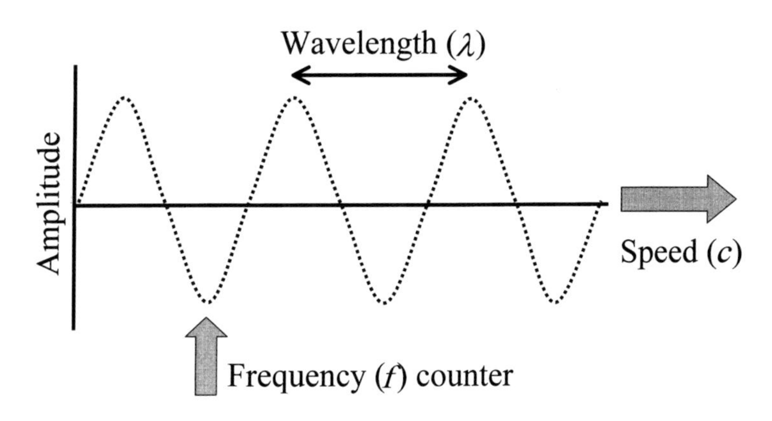

Fig. 120 *The characteristics of a wave. The wavelength, λ, corresponds to the distance over which the wave begins to repeat itself. The frequency, f, is the number of wave peaks (or troughs) that pass a given point per second. The unit of frequency is the hertz (Hz), named in honour of Heinrich Hertz, who discovered radio waves; 1Hz is one cycle per second.*

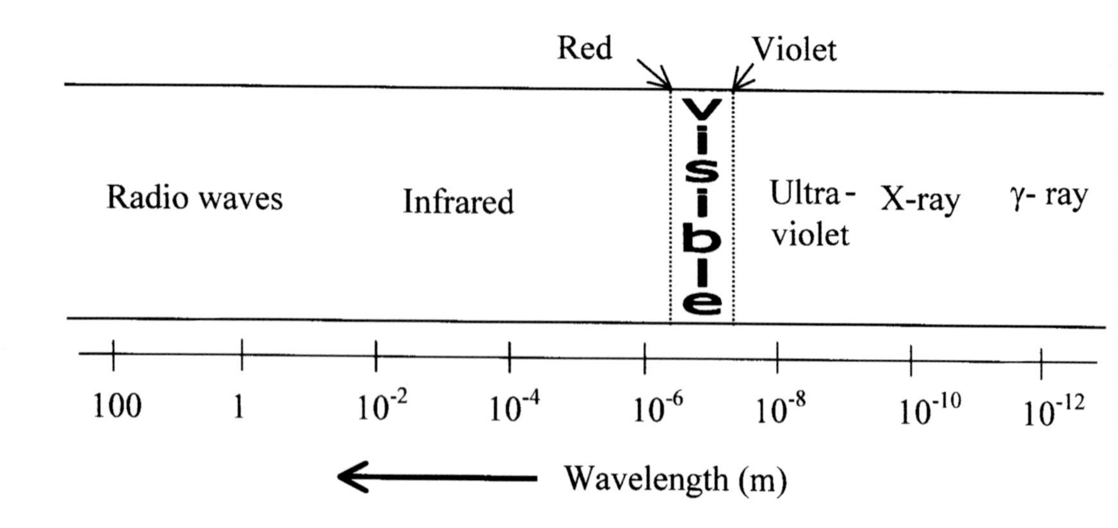

Fig. 121 *A schematic electromagnetic (em) spectrum. The x-axis indicates the wavelength, λ, in metres, which increases from right to left. The radiation that we see with our eyes (visible light) corresponds to just a small wavelength region (λ = 400–700nm, where 1nm = 10^{-9}m).*

propagate through space at the speed of light. The key distinction between the various forms of electromagnetic radiation is their wavelength. The wavelength is literally the length of a wave: the distance from one peak of the wave to the next . The progression across what is known as the electromagnetic spectrum, from gamma (γ) rays via X-rays, UV radiation, visible light and IR radiation to radio waves, corresponds to an increase in the wavelength of the radiation. A fundamental relationship between frequency, f, the number of wave cycles per second, and the wavelength, λ, is $\lambda f = c$, where c is the speed of light (3×10^8m/s).

Meteor Train Formation

As a meteoroid undergoes ablation in the Earth's upper atmosphere, the atoms that evaporate from its surface are ionized through collisions with molecules in the atmosphere. These ionizations result in the formation of a long, cylindrical train of plasma, composed of electrons and ions, trailing behind the meteoroid, and it is this plasma column that can interact with radio waves. If the degree of ionization – the ionization density – is sufficiently low, then a radio wave can penetrate into the plasma column without significant modification, but it will eventually be scattered, to move in a new direction of propagation, by interactions with individual electrons. Under these circumstances the meteor train is said to be **underdense**. If, on the other hand, the ionization density is sufficiently high, then the incident radio wave will be totally reflected by the plasma column, which behaves as if it were a metallic reflector, and the meteor train is called **overdense**. It has become standard practice to express the electron density as a so-called **line density**, which is literally the number of electrons per metre length of meteor train column. The changeover from the underdense to the overdense train condition occurs at an electron line density of about 10^{14} electrons per metre. Meteors with a maximum brightness greater than about magnitude +5 will typically produce overdense trains.

Since meteoroids enter the atmosphere at hypersonic speeds, the ionized meteor train is formed very rapidly. The initial radius of the train will depend upon the altitude at which ablation takes place. At an altitude of 85km the initial radius is about 0.5m, while at 115km it is more like 3m. Once the train has formed it begins to expand, the plasma column growing wider and wider. This steady increase in the width of the plasma column has important consequences for its interaction with any incident radio waves. Once the train's width is comparable in size to the wavelength of an incident radio wave (i.e. several metres across), then the radiation scattered from the near and far parts of the column will interact destructively, with wave peaks cancelling wave troughs, causing the signal strength of the scattered radio wave to become very small. The signal strength of any radio wave scattered by an underdense meteor train will decay exponentially as the train expands.

Figure 122 shows how the signal strength varies for a typical overdense radio echo. The very steep echo rise time (from S to P1) corresponds to the rapid formation of the meteor train by collisions between the atoms evaporated from the meteoroid and molecules in the atmosphere. Between P1 and P2 the electron density is high enough for the train to be overdense, even though at this point the train is expanding radially outwards. After P2 the echo strength begins to decline exponentially as electrons and ions in the plasma start to recombine into neutral atoms, and because of the gradual decrease in the electron density resulting from the continued radial expansion of the train into the surrounding atmosphere. At E on the curve the train has either expanded to the point where it

Fig. 122 *A schematic overdense radio echo profile. The diagram shows how the strength of the radio signal strength recorded by the receiver varies with time. An underdense radio echo would not show the plateau between P1 and P2, but would simply begin to decay after P1. A typical underdense echo lasts for just a fraction of a second; an overdense trail can last for many seconds.*

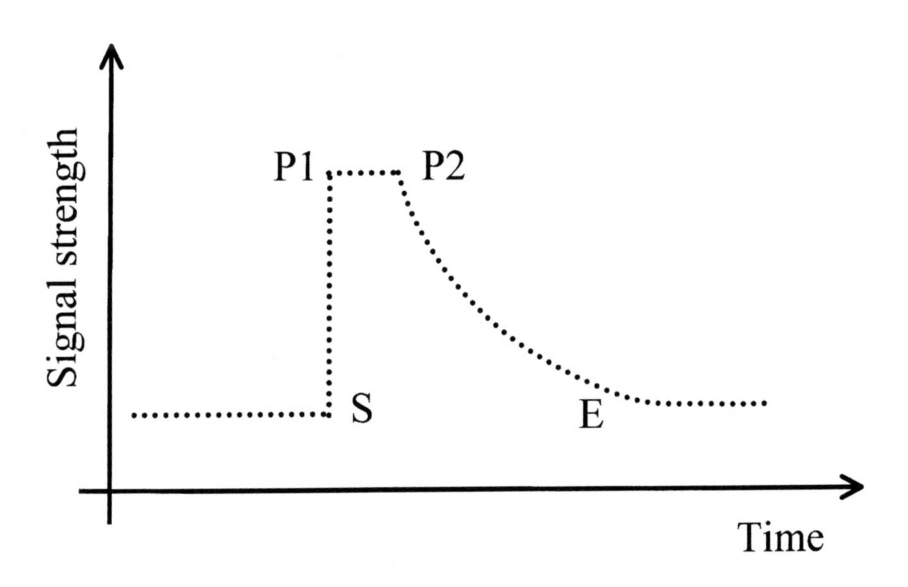

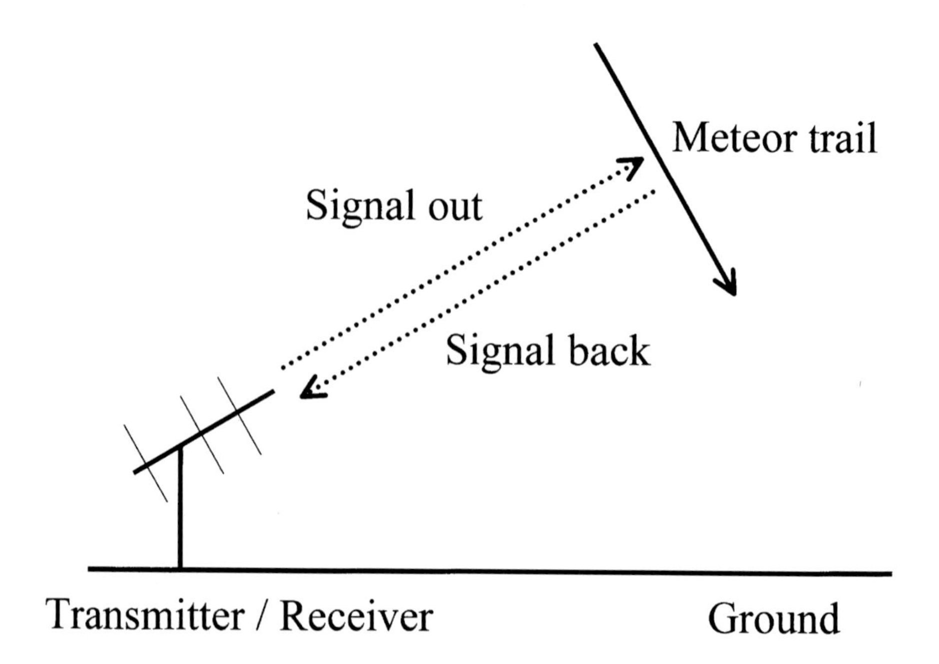

Fig. 123 (above) *The arrangement of a backscatter system for detecting meteors. Since the transmitter and receiver are at the same location, only those meteor trains whose paths are at right angles to the direction of the transmitter radio pulse will be detected.*

Fig. 124 (below) *The arrangement of a forward-scatter system for detecting meteors. The separation between the transmitter and receiver is typically many hundreds, if not a thousand or so kilometres. Such systems are most sensitive to the detection of meteors that appear midway between the transmitter and the receiver.*

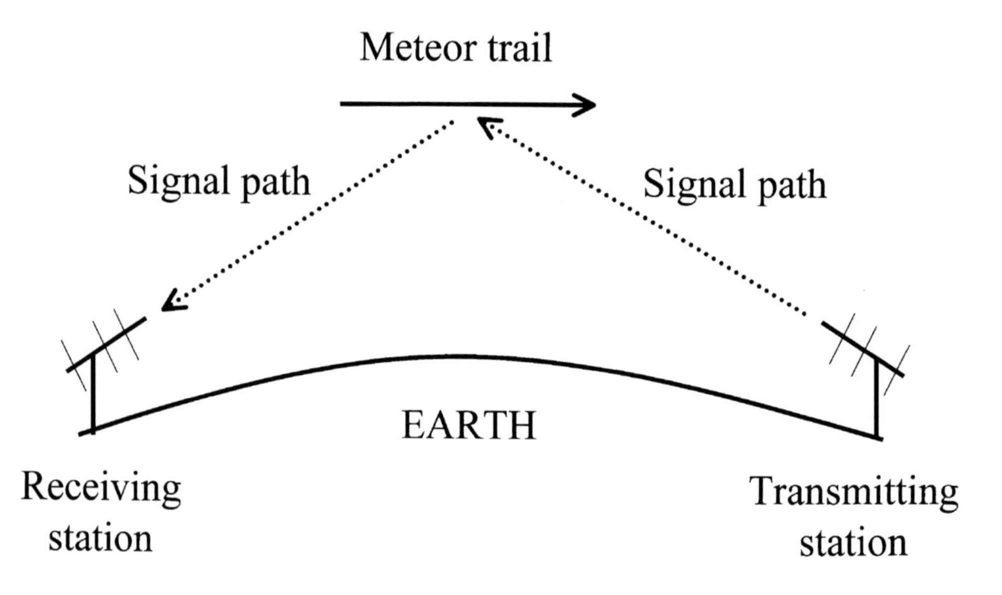

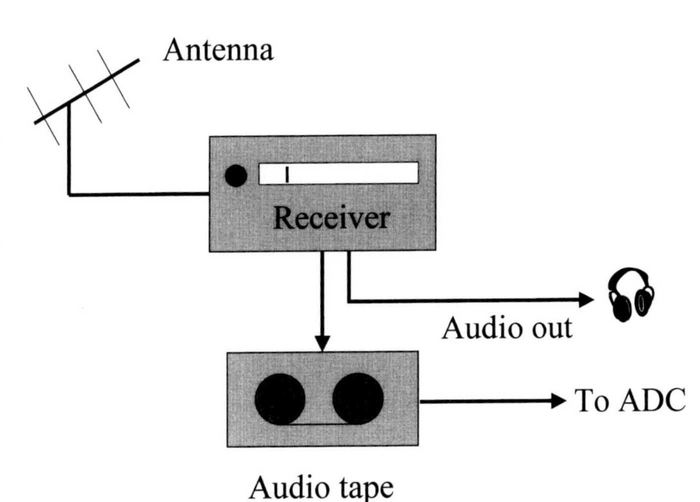

Fig. 125 *The layout of the main components of a basic forward-scatter radio meteor detection system. The receiver can be any commercially brought radio, and (to begin with) the antenna need be no more than a simple wire or V-shaped dipole antenna. The output from the audio tape recorder connects to an analogue-to-digital (ADC) converter, which enables computer monitoring of the signal to be performed.*

is comparable in size to the wavelength of the transmitted radio wave (thus allowing destructive interference to occur), or all of the electrons responsible for scattering the transmitted radio wave have become reattached to ionized atoms.

Forward and Backscatter

There are essentially two methods that can be applied to the study of meteors at radio wavelengths: the so-called **forward-scatter** and **backscatter** techniques. The two methods differ according to the transmitter wavelength used and the location of the receiver with respect to the transmitter. Figure 123 shows a typical arrangement for a backscatter system in which the transmitter and receiver are at the same geographical location, and the transmitted radio wave interacts with the meteor's ionization column simply by being reflected straight back along the same path. Figure 124 shows a typical arrangement for a forward-scatter system. Here the transmitted signal is reflected by the meteor's ionization column at an angle, and the transmitter and receiver are not at the same geographical location – indeed, they are usually separated by many hundreds of kilometres, and the transmitter is over the horizon of the receiver station.

Most professional radio-meteor studies have been carried out at wavelengths in the range from about 5 to 15m (corresponding to frequencies of between 60 and 20MHz, respectively). Such studies, however, usually require the design and construction of specialized instruments. Fortunately, basic meteor monitoring can be performed with much less complex equipment than is used by professional astronomers. And by far the simplest way to study meteors at radio wavelengths is not to construct a special transmitter system, but to sample broadcasts from distant radio stations.

A Basic Forward-Scatter System

An elementary forward-scatter meteor observing station can be put together without having to buy any expensive equipment. A reasonably good radio receiver is required, but it need be no more exotic than a standard car radio. Since the radio will have to be tuned to a non-local radio station, its only essential feature is digital tuning. The antenna need only be a simple car antenna or one of the standard indoor, V-shaped dipole antennas that are often supplied with small portable television sets. Figure 125 shows the layout for a basic system.

The essence of forward-scatter meteor detection is to tune the receiver to an FM radio station that does not provide a direct, line-of-sight signal. The receiver, for example, can be shielded from receiving a direct signal by intervening hills and mountains, or the station can be over the horizon. The main point of this shielding is that the only time that a receiver should be able to record a signal from the station is when it is reflected by the ionization column of a meteor. For forward-scatter detection, without 'natural' shielding, the transmitter should be located about 1,000km from the receiver.

Ideally the station transmitter to tune to for forward-scatter work should be either due east or due west of the receiver. In the UK this essentially means that you will need to identify an Eastern European transmitter (I explain why below). Data on the location and transmitting frequencies of all the world's radio stations can be found in the *World Radio TV Handbook* (http://www.wrth.com). A copy of this handbook will probably be available for reference at most large libraries. For observers in North America, a list of radio station locations and transmitting frequencies can found in the *FM Atlas* (http://members.aol.com/fmatlas/home.html).

Having identified a suitable distant radio station, tune the receiver to the chosen frequency. At this point, all that you should hear is a steady background hiss. If you can hear a constant signal (voice or music) then the transmitter is too close for useful radio meteor work, and another, more distant radio station will have to be selected. When a meteoroid produces an ionization train in the right part of the sky for forward scatter, a brief signal will be heard over the background hiss. The signal will typically last just a fraction of a second and makes a sound called a 'ping'. Overdense trains, from larger meteoroids, might produce several seconds of clear transmission enabling a few words or bars of music to be heard. A forward-scatter radio system has been deployed at the Marshall Space Flight Center in Huntsville, Alabama, where a typical meteor echo sound can be downloaded and listened to. The Marshall website also has a live audio stream from their receiver, at http://www.spaceweather.com/glossary/nasameteorradar.html/. The Marshall system is tuned to a frequency of 67.25MHz (λ = 4.46m), and uses a TV transmitter to produce the forward-scatter signal. The Jordanian Astronomical Society (JAS) has prepared a very readable introductory website relating to forward-scatter observation: http://www.jas.org.jo/radio.html/.

The main function of any forward-scatter monitoring station is to provide information on the arrival rate of meteors. The rate will vary according to the time of day and the time of year. If you have an indoor aerial then perhaps five to ten meteor 'pings' per hour might be heard when a meteor shower, with a well placed radiant, is close to its maximum activity. This number of detections, however, can be greatly increased by using a more sensitive outdoor aerial (*see below*).

A major drawback with forward-scatter observations is that no clear distinction can be made between sporadic and shower meteors. The sporadic background of meteors varies systematically over the course of the day (just as with visual observations, as described in Chapter 2), with a maximum at around 4am local time and a minimum around 4pm. Since the geometry of the forward-scatter interaction means that only those meteor trains that are perpendicular to the line joining the transmitter and receiver will be detected, the number of shower meteors recorded per day will vary according to the position of the shower's radiant in the sky. The directional sensitivity of a forward-scatter radio system is described by the so-called variability function. Shower meteors will be best placed for producing forward-scatter signals when the radiant is positioned at 90° to the transmitter–receiver line. Since a

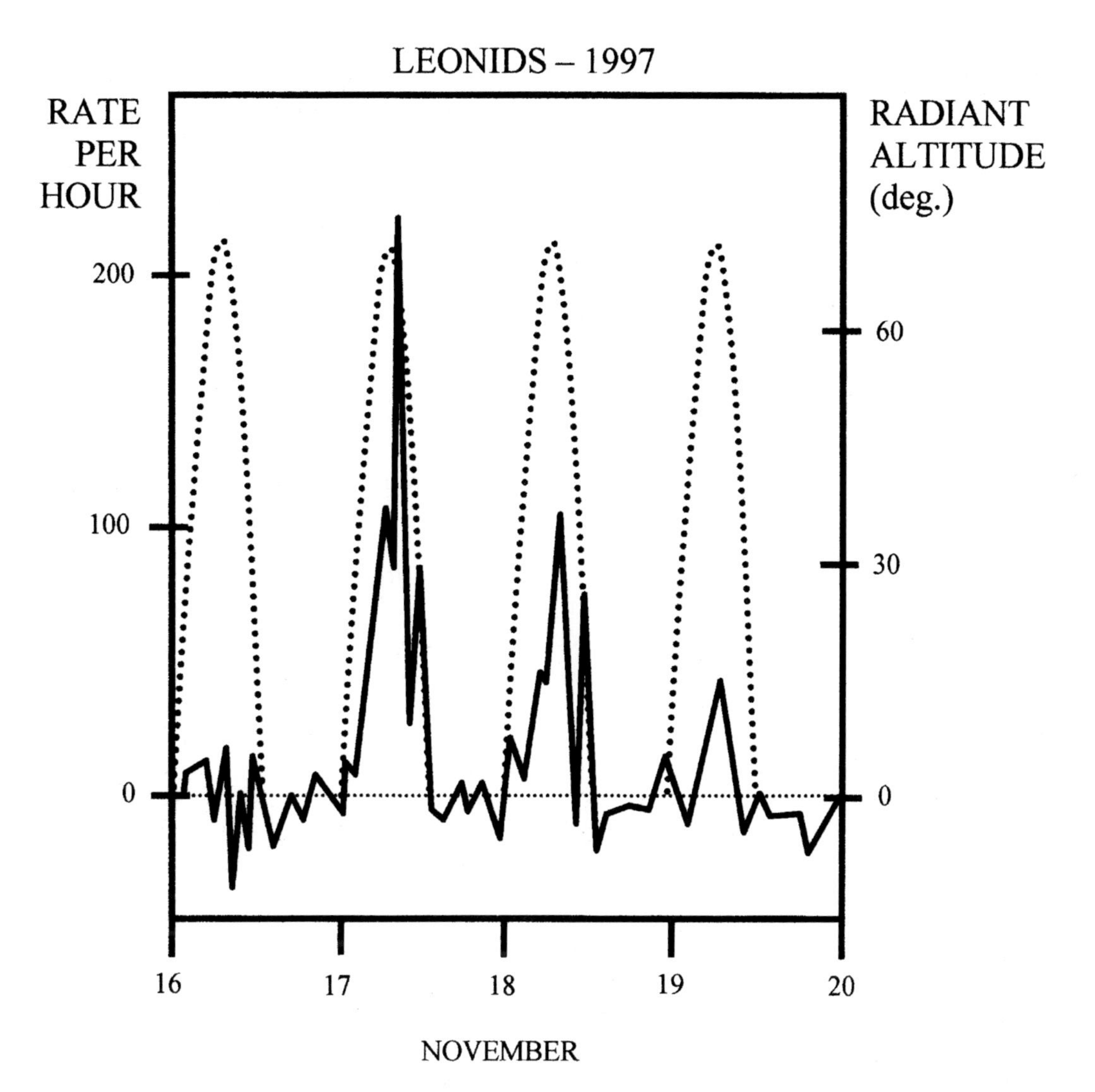

Fig. 126 *Leonid meteor rates during the time interval from 16 to 20 November 1997. The y-axis shows the number of radio echoes recorded per hour on the left, and the Leonid radiant altitude on the right. An average background count has been subtracted from the recorded number per hour (solid line). The corrected counts should therefore be (mostly) Leonid meteors and not sporadic meteors. The dotted lines show the radiant's altitude for the observing site (Greenbelt, Maryland). When the radiant altitude is less than zero, the Leonid meteor count should also be zero. The peak Leonid meteor rate occurred at 8am (local time) on 17 November. (Based upon data from A. Mallama and F. Espenak,* Publications of the Astronomical Society of the Pacific, *vol. 111, pp. 359–363, 1999)*

shower radiant will always culminate towards due south, the optimal orientation for the transmitter–receiver line will be east–west. The variability function will be close to zero when the radiant is directly aligned with the transmitter–receiver line. As the orientation of the radiant's position changes with the rotation of the Earth, so the variability function will also change. In practice, what this means is that even if the meteor rate from a shower radiant is constant with time, the number of meteors actually recorded will vary, with the greatest number of meteors being recorded when the radiant is perpendicular to the transmitter–receiver line. In this manner one can certainly monitor meteor shower activity, but it is very difficult to convert the observed forward-scatter detection rate to an actual zenithal hourly rate for the shower.

Figure 126 shows how the radio meteor rate varied with time and radiant altitude for the 1997 Leonid meteor shower. Clearly, when the radiant is below the observer's horizon, the shower meteor count will be zero. When the radiant culminates (i.e. at its maximum altitude and due south), the shower meteor count will be at its greatest for a system in which the transmitter–receiver line is orientated in the east–west direction.

Improving the System

The efficiency of the forward-scatter system described above can be improved upon by adding a Yagi antenna and automating the meteor detection process. The first improvement will enable the detection of more meteors by providing a stronger, more directional signal, and the second will ensure that all the meteors are counted (after a while it is easy for the human listener to lose concentration and miss events). In addition, by automating the meteor detection and counting, the system can be left to run 24 hours a day, seven days a week, 52 weeks of a year, which will result in the collection of good data on the yearly variation of meteor activity. With such data one should be able to discern the seasonal variability of the sporadic meteor background rate (*see* Figure 26, p. 30) and identify the times at which the annual meteor showers occur.

Constructing a Yagi Antenna

The Yagi antenna (designed in the 1930s by Japanese radio engineer Hidetsugu Yagi) provides both good signal gain and directionality. The antenna consists of a reflector, a driven folded-dipole element and a series of directors. The lengths and spacings of the various antenna elements are determined by the wavelength at which the system is going to work (i.e. the wavelength of the distant radio station to be monitored). The simplest Yagi antenna has just three elements, and instructions for making such an antenna can be found at the educational *Sky Scan* website: http://www.skyscan.ca/3ElementYagi.htm/. In addition, Ilkka Yrjölä has provided plans for a four-element Yagi antenna, usable over the frequency range 87–92 MHz, on his website at http://www.kolumbus.fi/oh5iy/msobs/msobs.html/. The lengths of the various elements in terms of the wavelength (λ) of the chosen radio station are:

Reflector length: 0.495λ
Driven element length: 0.473λ
First director element length: 0.440λ
Second director element length: 0.435λ
Third director element length: 0.430λ

The spacings between the various elements are:

Reflector to driven element: 0.125λ
Driven element to first director: 0.125λ
Separation between directors: 0.250λ

So, for example, if the chosen frequency of the radio station to be monitored is 95MHz, then the wavelength to design for is

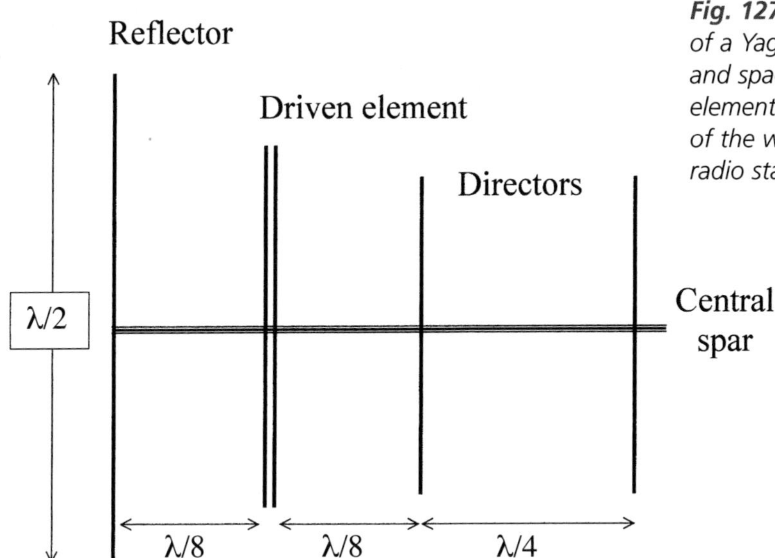

Fig. 127 *A schematic diagram of a Yagi antenna. The lengths and spacings of all the elements are simple fractions of the wavelength, λ, of the radio station to be monitored.*

$$\lambda = \frac{3 \times 10^{8}\ (\text{m/s})}{95 \times 10^{8}\text{Hz}} = 3.16\text{m}$$

Consequently, the reflector element should be 1.563m long, and the driven element 1.494m long (these lengths should be accurate to the nearest millimetre or two). The first director element should be 1.390m long, and the second director element 1.374m long. The spacing between the reflector and the driven element should be set to 0.395m, as should the spacing between the driven element and the first director. The subsequent director elements should be separated by intervals of 0.789m.

It is a good idea to include as many director elements as can be accommodated in the space available for the antenna. But bear in mind that the more elements on the Yagi, the longer it will be and the more substantial must be the mount. A three-element Yagi is a good first antenna to build. The central spar can be a length of wood or plastic pipe. The reflector and driver elements can be made from single-core copper wire cut into appropriate lengths. Use waterproof glue to attach the copper wire strips for the reflector and director elements to wooden dowels, and then secure the dowels at their appropriate locations, lengthwise across the central pole. The driven element is constructed from 300Ω impedance twin-lead (also called ladder line) cable, which is readily available from most commercial electronics supply shops. Cut the twin lead to the correct length for the frequency of the radio station chosen, and glue it to a piece of wood attached across the central spar. Carefully make a cut, at the centre position, in one of the twin lead strands, and attach a coaxial cable to the exposed wire ends (*see* Figure 128). The coaxial cable is then connected to the receiver antenna socket.

Remember that since the Yagi antenna will be mounted outdoors, it will have to endure occasional strong winds and poor weather. So it is worth spending some time and money on constructing a good, sturdy antenna – it should not flex under its own weight or in the wind – and make sure that it is securely mounted to a mast or building.

Once the antenna has been constructed, then it should be mounted so as to point in the

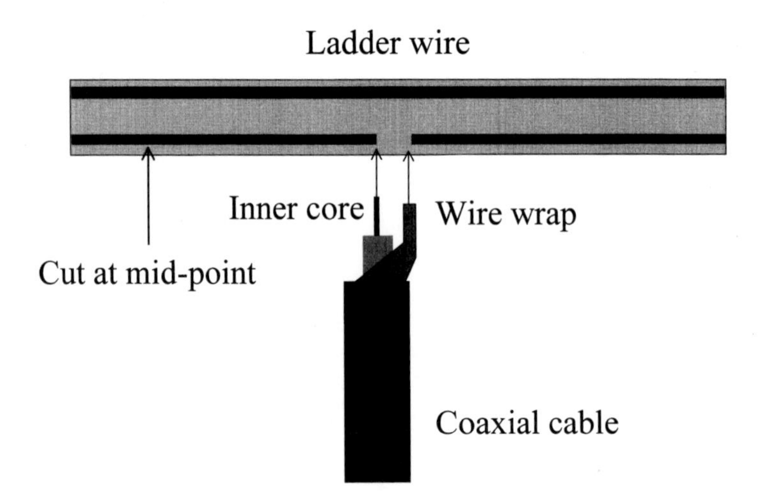

Fig. 128 *The construction of driven elements for a Yagi antenna. Only one of the ladder lead wires is cut, and the coaxial cable is run to the receiver input.*

direction of the radio station to be monitored. Use a map and compass to determine the bearing, and then adjust the antenna so that it is pointing within a few degrees of the required position. The best elevation for the reception of meteor echoes will have to be found by experimentation. Begin with the antenna in a horizontal position and increase the elevation according to the general level of background noise and the best forward-scatter reception. The International Meteor Organization's website provides a freeware program called *Yagimax* that can be used to calculate the expected performance characteristics of any Yagi antenna design. The program (MS-DOS only) can be downloaded from http://www.imo.net/radio/software.html/.

Automated Meteor Detection

To automate meteor detection you will need to have access to a computer, an analogue-to-digital converter and some dedicated software. The computer need not be state of the art, but should be able to run under the Windows environment. The analogue-to-digital converter (ADC) is the interface between the computer and the audio output from the receiver (*see* Figure 125). The ADC essentially converts the continuously varying electrical output from the receiver into a digital signal that the computer software can read and interpret.

Figure 129 shows the circuit diagram for an ADC designed by Antonio Picar of the Central University of Venezuela. The circuit utilizes the National Semiconductor ADC0831 integrated chip. You will need to consult the manual for your computer in order to identify which of the RS 232 port pins should be used; you will also need to identify the clock (CLK), the request to send (RTS) and the data terminal ready (DTR) pins. The variable resistor (R5), connected to the V_{ref} pin of the ADC0831 chip, allows the maximum sampling voltage to be adjusted. The audio output from the receiver is connected to V_{in}.

The components for the ADC are all readily available from any electronics supply store. An alternative design for an ADC can be found at the website developed by Pierre Terrier: http://radio.meteor.free.fr/us/main.html/. This website is especially recommended for first-time enthusiasts since it offers a step-by-step construction guide for an ADC. The website also supports a freeware program called *Meteor* (which runs under MS-DOS) that will detect, store and register meteor echoes – for those not experienced in computer programming, this is a great boon. Radio-Sky

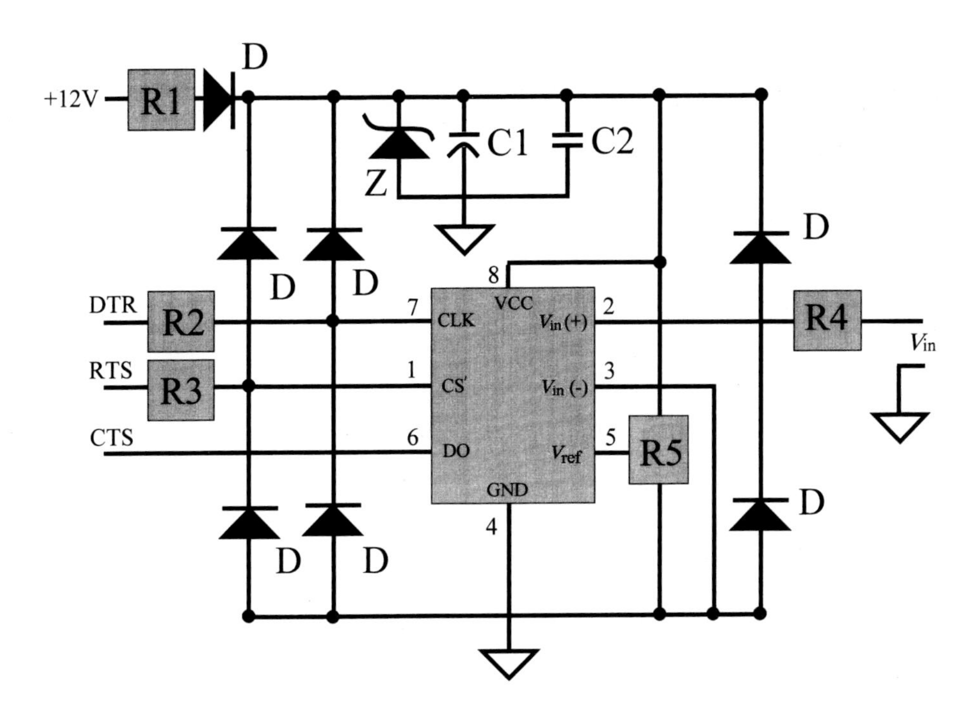

***Fig. 129** ADC circuit diagram. Solid dots indicate wire connections. Components: R1 = 1kΩ, R2 = 30kΩ, R3 = 30kΩ, R4 = 50kΩ, R5 = 10kΩ (variable); C1 =10µF, C2 = 100nF; Z = 5.1V Zener diode, D = standard signal diodes (e.g. IN4148); IC = National Semiconductor ADC0831; RS 232 connector cable and plug. (Based upon the design by Antonio Martinez Picar described in WGN, vol. 31 (5), pp. 139–144, 2003)*

Publishing in Hawaii sells an ADC kit (the MAX186 ADC) that attaches to a computer's parallel port, and the design is supported by a freeware program called *Radio-SkyPipe*. This program will also work with any 16-bit sound card and will automatically detect, store and register meteor echoes. The web page for further details on the MAX 186 ADC and *Radio-SkyPipe* program is http://www.radiosky.com/skypipeishere.html/.

Appendix

The Quadrantids

Duration	Maximum ($\lambda_{\odot}$)	ZHR	r	RA	Dec.
1 Jan. – 5 Jan	283.3°	120	2.2	230°	+49°

The possible existence of a meteor shower in early January, with a radiant in the now obsolete constellation Quadrans Muralis (the Mural Quadrant), was first noted in the early 1800s. The shower now shows a rapid rise to maximum, with high meteor rates lasting for just a few hours (*see* Figure 82). The Quadrantid ZHR at maximum is one of the highest among the principal annual meteor showers, but the relatively low initial speed of Quadrantid meteoroids (39km/s) dictates that most of the meteors are faint. Given the poor observing conditions that typically prevail when the shower is active, the Quadrantids are well suited for further study.

Although the mean stream orbit has a high inclination of some 73° at the present time, it has been closer to the ecliptic, with a much smaller inclination, in the past. Computer modelling of the stream indicates that some 1,500 years ago the inclination was as low as 13° and the perihelion distance as small as 0.1AU, nearly ten times smaller than the stream's present perihelion distance. These variations in the stream's orbital characteristics are driven by gravitation perturbations by the planet Jupiter. The computer simulations also indicate that in about 400 years' time the stream's perihelion distance will be greater than 1AU, and consequently the stream will no longer intercept the Earth's orbit. Once that happens, then, just like the constellation after which the shower is named, the Quadrantids will be no more. The parent comet of the Quadrantid meteoroid stream has recently

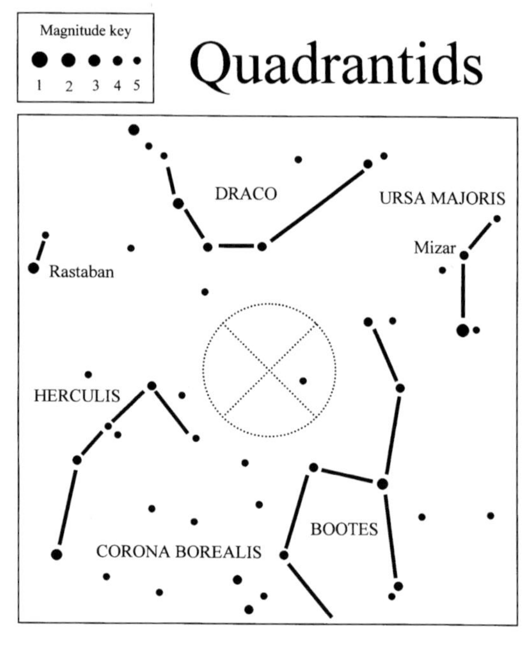

Fig. A1 *The sky location of the Quadrantid radiant on the night of the shower's maximum (4 January). The field of view is 40° × 40°.*

been identified by Peter Jenniskens (NASA Ames Research Center) as the object designated 2003 EH1. Although this is a minor planet designation, it is thought that 2003 EH1 is actually the remnant of an old and now inactive cometary nucleus.

The Lyrids

Duration	Maximum ($\lambda_{\odot}$)	ZHR	r	RA	Dec.
16 Apr. – 25 Apr.	32.3°	20	2.9	271°	+34°

The Lyrids burst onto the contemporary meteor scene on the night of 19 April 1803, when the shower delivered a dramatic outburst of meteors witnessed from across the eastern United States. Rates of some 650 meteors per hour were seen at the time of maximum activity. Outbursts have also been recorded in 1922 and 1982. More typically, however, the shower produces a few tens of meteors per hour at maximum. The quite high atmospheric initial speed of Lyrid meteoroids (48km/s) makes Lyrid meteors quite bright and distinct. The shower, which is associated with the long-period comet Thatcher, can be traced as far back as 687BC, making it the oldest of the known annual meteor showers.

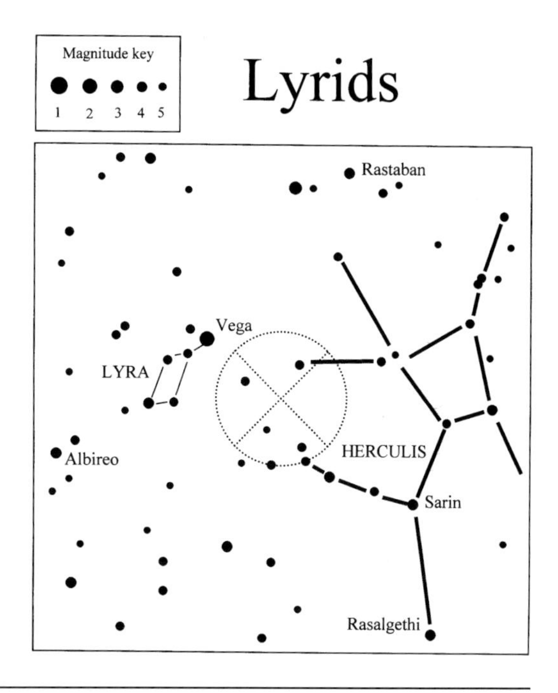

Fig. A2 The sky location of the Lyrid radiant on the night of the shower's maximum (22 April). The field of view is 40° x 40°.

The Eta Aquarids

Duration	Maximum ($\lambda_{\odot}$)	ZHR	r	RA	Dec.
19 Apr. – 28 May	45.5°	60	2.4	338°	–02°

The Eta Aquarid meteor shower is associated with Halley's comet (*see* Figure 36). The meteoroids, just like the parent comet, move around the Sun in a retrograde fashion and consequently they encounter the Earth with a high initial speed, of 65km/s; the meteors are correspondingly swift and bright, and they often leave short-lived persistent trains. The Eta Aquarids are twinned with the Orionids, which the Earth encounters in late October each year (*see* Figure 39).

Because the radiant is in the southerly constellation of Aquarius, the shower is not particularly well placed for viewing in the

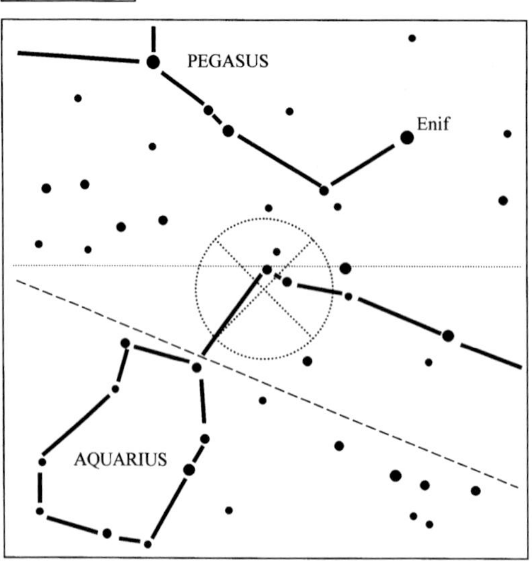

northern hemisphere, with the radiant rising only in the pre-dawn hours. Activity from the Eta Aquarids can be traced back as far as 74BC, but there is no evidence of the shower ever having produced any spectacular storms or outbursts.

Fig. A3 *The sky location of the Eta Aquarid radiant on the night of the shower's maximum (5 May). The horizontal dotted line is the celestial equator, and the dashed diagonal line is the ecliptic. The field of view is 40° × 40°.*

The Perseids

Duration	Maximum ($\lambda_{\odot}$)	ZHR	r	RA	Dec.
17 Jul. – 24 Aug.	140.0°	100	2.6	46°	+58°

The Perseid meteor shower is arguably the best known and most observed of all the annual meteor showers. Active throughout the month of August (*see* Figure 35), the shower occurs when the weather is typically fine and many people are actively observing. The meteoroids are derived from comet Swift–Tuttle and enter the Earth's atmosphere at initial speeds of about 60km/s. Perseid meteors are bright and swift, and the shower can deliver spectacular fireballs. The peak rate is consistently high, at around 100 meteors per hour (*see* Figure 82), although outbursts have been known. The

Magnitude key
1 2 3 4 5

Perseids

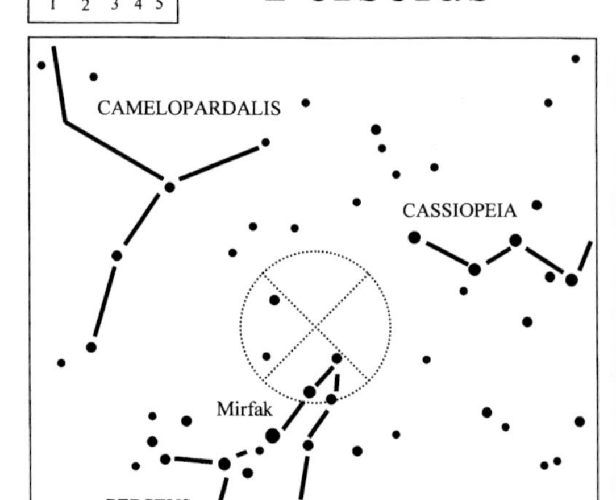

Fig. A4 *Sky location of the Perseid radiant on the night of shower maximum (12 August). The position of the radiant from 20 July to 20 August is shown in Figure 35. The field of view is 40° × 40°.*

meteor rates were particularly high in the early 1990s, around the time of the parent comet's last return to perihelion (December 1992).

The earliest account of the Perseid meteor shower is from AD36, when Chinese records indicate that 'more than a hundred meteors flew thither in the morning'. In more recent centuries the Perseids have been known as 'the Tears of St Lawrence' – a reference to the fact that the shower's maximum falls close to 10 August, Lawrence's feast day. In many ways the Perseids are the 'old faithful' of meteor showers, with maximum rates and an activity profile that are remarkably consistent from one year to the next. During the early 1990s the Perseids developed a transient early maximum, appearing about a day before the 'traditional' maximum on 12 August. This new peak, which since 1996 has all but vanished, was the result of the Earth encountering material ejected by comet Swift–Tuttle during its 1862 perihelion passage. In 1993, ESA's Olympus telecommunications satellite was put out of action by the impact of a Perseid meteoroid, making it the first meteoroid-induced casualty of the space era.

The Orionids

Duration	Maximum ($\lambda_{\odot}$)	ZHR	r	RA	Dec.
2 Oct. – 7 Nov.	208.0°	25	2.5	95°	+16°

The Orionid meteor shower is the twin of the Eta Aquarids. The shower has a complex radiant structure which has been interpreted as indicating a distinct filamentary structure to the stream. The activity profile of the Orionids is highly variable, and the shower is most definitely a strong candidate for continued monitoring and study. Orionid meteors are swift, having an initial speed of some 66km/s, and a high proportion of them are accompanied by short-lived persistent trains.

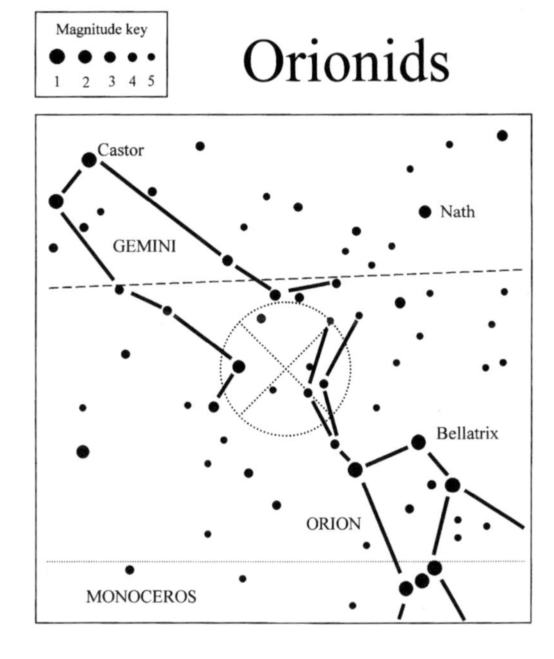

Fig. A5 The sky location of the Orionid radiant on the night of the shower's maximum (21 October). The horizontal dotted line is the celestial equator, and the dashed diagonal line is the ecliptic. The field of view is 40° × 40°.

The Northern and Southern Taurids

Two branches of the long-duration Taurid shower are recognized. The northern branch reaches a maximum around 5 November, while the southern branch peaks around 12 November.

Duration	Maximum ($\lambda_{\odot}$)	ZHR	r	RA	Dec.
1 Oct. – 25 Nov.	223.0°	15	2.3	52°	+13°
1 Oct. – 25 Nov.	230.0°	15	2.3	58°	+22°

The meteoroids from both branches of the Taurid shower are derived from comet Encke and form part of the Taurid complex, an extensive group of orbitally related comets, Apollo asteroids and dust trails. Taurid meteoroids enter the Earth's atmosphere at about 28km/s, and the shower is known for occasionally producing exceptionally bright fireballs.

Also associated with comet Encke are the Beta Taurid and Zeta Perseid meteoroid streams. Both these streams produce daytime meteor showers, in mid to late June of each year, and are accessible to study only via radio and radar techniques.

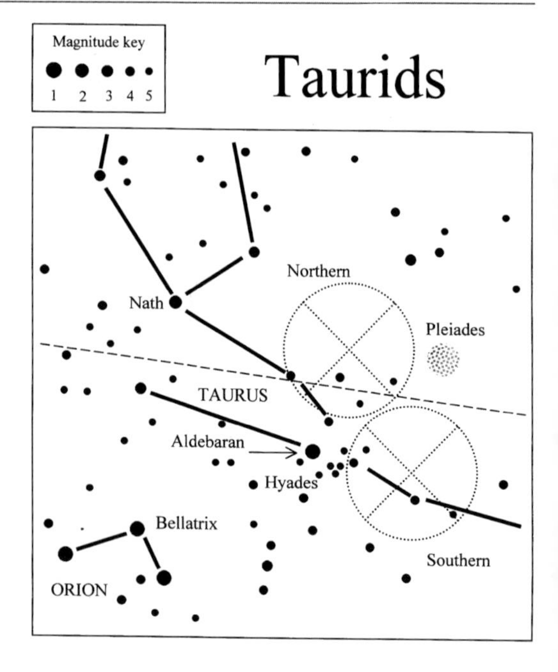

Fig. A6 The sky location of the Taurid radiants on the night of the shower's maximum. The Southern Taurids peak on 5 November, while the Northern Taurids peak on 12 November. The dashed diagonal line is the ecliptic. The field of view is 40° × 40°.

The Leonids

Duration	Maximum ($\lambda_{\odot}$)	ZHR	r	RA	Dec.
14 Nov. – 21 Nov.	235.3°	20	2.5	153°	+22°

The Leonid meteor shower is famous for producing spectacular meteor storms (*see* Figure 103). Indeed, the last such storm/outburst cycle has only recently finished. The first record of a Leonid storm dates back to 13 October AD902. On that date observers in Sicily reported seeing 'small star-like fires' in the sky, and an Arabic chronicle records that 'stars sprinkled all over the atmosphere in Egypt'.

Leonid meteors are very swift, having an initial speed of about 70km/s; this high speed dictates that many Leonids are accompanied by persistent trains (*see* Figures 53 and 54). Outside of the times of enhanced activity, controlled by the 33.3-year periodic returns of the parent comet, Tempel–Tuttle, to perihelion, the shower produces only a modest maximum of perhaps 20 meteors per hour. The existence of distinct 'streamlets' within the Leonid stream requires more observational analysis and investigation (*see* Figure 83).

An analysis of the times at which Leonid storms have historically occurred indicates that they are most likely to be seen when the Earth passes through the region just outside the orbit of comet Tempel–Tuttle shortly after it has passed through its nodal point. By carefully studying the evolution of the comet's orbital characteristics, Donald Yeomans and colleagues at the Jet Propulsion Laboratory (JPL) in Pasadena, California, have found that the conditions for producing strong Leonid storms will probably not occur again until at least the 2098 return of comet Tempel–Tuttle. It would seem, therefore, that the most recent Leonid storm cycle was the last one for several generations.

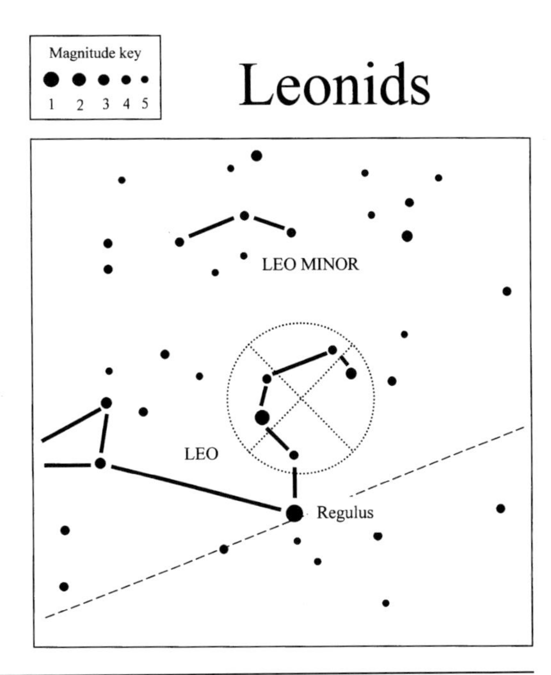

Fig. A7 *The sky location of the Leonid radiant on the night of the shower's maximum (17 November). The dashed diagonal line is the ecliptic. The field of view is 40° × 40°.*

The Geminids

Duration	Maximum ($\lambda_{\odot}$)	ZHR	r	RA	Dec.
7 Dec. – 17 Dec.	262.2°	120	2.3	112°	+33°

The Geminids produce the strongest and most distinct display of all the annual meteor showers. The meteors are slow, having initial speeds of 35km/s, but they can be very bright. The Geminids are also distinguished by their ability to display a rapid flickering phenomenon, probably related to meteoroid rotation (*see* Figure 115). Meteors from the Geminid shower were first noted in the early 1860s, and the shower's activity profile is unusual in that it displays a slow rise to maximum and then a rapid fall (*see* Figure 30). In addition, there is a higher proportion of faint

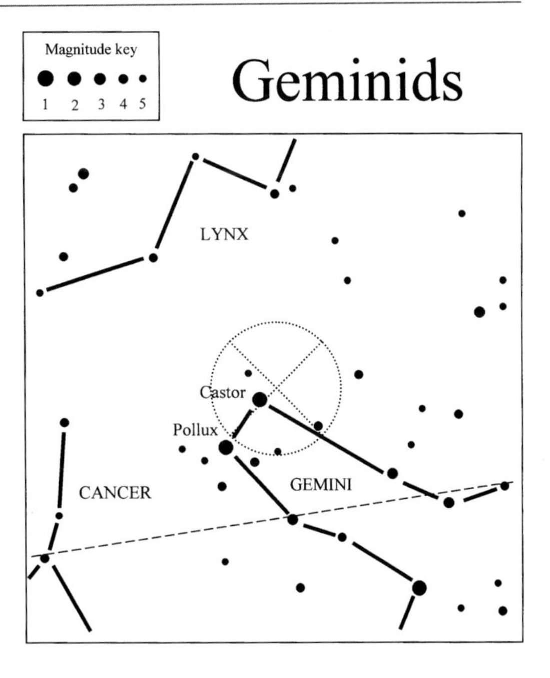

Fig. A8 *Sky location of the Geminid radiant on the night of shower maximum (14 December). The dashed line is the ecliptic. The field of view is 40° × 40°.*

meteors to bright ones than in other showers before the time of shower maximum, with the reverse effect occurring afterwards. This indicates that meteoroid mass sorting, whereby less massive meteoroids preferentially move into orbits that begin to intercept the Earth's before more massive meteoroids do, has taken place within the stream, and James Jones (University of Western Ontario, Canada) has used computer modelling of this effect to deduce a stream age of about 5,000 years. The parent object of the Geminid stream is the Apollo asteroid Phaethon, an object that is, in fact, an old, inactive cometary nucleus.

Glossary

ablation
Any process that removes material from the surface of a meteoroid.

achondrite
A stony meteorite, lacking *chondrules*, that crystallized from molten rock.

aphelion
The point at which an object, be it a planet, a comet, an asteroid or a meteoroid, in orbit about the Sun is at its greatest distance from the Sun.

argument of perihelion (w)
The angle between the perihelion point and the ascending node in the plane of the orbit.

asteroid
A minor planet composed predominantly of silicates and metals. The main-belt asteroids move along circular orbits about the Sun in the region between Mars and Jupiter. Some have highly eccentric orbits that allow them to periodically cross the Earth's orbit. These near-Earth asteroids (NEAs) can potentially impact the Earth.

Astronomical Unit (AU)
The mean distance of the Earth from the Sun (the *semi-major axis* of the Earth's orbit). 1AU = 1.496×10^8km.

azimuth
The direction of a celestial object, measured in degrees clockwise from north.

backscatter
A method of detecting meteor trails via the back reflection of a radar signal.

basalt
A common volcanic rock composed of pyroxene and plagioclase.

breccia
A rock or meteorite composed of many different rock fragments – often produced during impact events.

carbonaceous chondrite
A stony meteorite with a composition that closely matches that of the Sun (excluding hydrogen and helium). Such meteorites have suffered very little heat alteration, but typically show signs of having reacted chemically with water.

chondrite
An undifferentiated stony meteorite characterized by the presence of *chondrules*.

chondrule
A millimetre-sized spherule of rapidly cooled silicate material. Chondrules are the characteristic components of chondritic meteorites.

comet
A body made of dust and ice orbiting the Sun.

Cometary nuclei are typically a few kilometres across and are predominantly composed of water ice. The **short-period comets** tend to orbit the Sun in the plane of the ecliptic and are fragments of disrupted Kuiper Belt objects. **Long-period comets** (those having periods greater than 200 years) have highly elliptical orbits that carry them well out of the plane of the ecliptic. The long-period comets originate in the Oort Cloud.

culmination
The moment when a heavenly object is due south and at its highest apparent elevation for an observer.

dark adapted
Describing the condition of the human eye under low light levels when the rod receptors on the eye's retina are at their most sensitive.

dark flight
The final portion of a meteorite's flight before it hits the ground, after surface heating and ablation have stopped and no more light is being generated.

declination (dec.)
The north–south coordinate on the celestial sphere (similar to terrestrial latitude) measured from the celestial equator towards either celestial pole.

differentiation
The process by which an initially homogenous object (e.g. an asteroid or planet) becomes internally stratified.

eccentricity
The ratio of the distance from the centre of an ellipse (or an elliptical orbit) to the ellipse's focal point divided by the *semi-major axis*. It is a measure of how much an ellipse departs from a circle.

ecliptic
The apparent path of the Sun across the celestial sphere. The plane of the ecliptic is thus the plane of the Earth's orbit about the Sun.

electromagnetic radiation
Radiation consisting of waves with electric and magnetic components. All such waves travel at the speed of light.

electrophonic sound (simultaneous sounds)
Crackling and buzzing sounds heard at the same time that a fireball is observed, believed to be generated by the conversion of very long wavelength (*VLF*) electromagnetic radiation into audible sounds.

escape speed
The speed that a body must attain in order to break away from the gravitational attraction of another body.

fall
A meteorite with a known fall date.

find
A meteorite that was not observed to fall.

fireball
A meteor brighter than magnitude –4.7.

focal ratio (f-number)
A lens rating determined by dividing the lens's focal length by its diameter.

forward scatter
A method of detecting meteor trails by the forward reflection of a radio signal.

frequency
The number of cycles of a wave per second.

fusion crust
The thin, dark, iron-rich, glassy exterior of a

meteorite formed when its surface is heated by *ablation* during its passage through the atmosphere.

Global Positioning System (GPS)
A navigation system enabling precise timings and measurements of latitude and longitude to be obtained by receiving signals from a fleet of Earth orbiting spacecraft via a portable receiver.

great circle
Any of the circles of maximum diameter that can be drawn on the surface of a sphere.

inclination
The angle between the plane of an orbit and the plane of the ecliptic.

initial speed
The speed of a meteoroid relative to the Earth when it first enters the Earth's atmosphere.

interference
The superposition of two or more waves, resulting in either amplitude enhancements or cancellations (destructive interference).

ionization
The process by which an atom loses or gains electrons. A gas that has been ionized contains both free electrons and ions (positively charged atoms). A **plasma** is an electrically neutral ionized gas – that is, in a plasma there are equal numbers of electrons and ions.

iron
A meteorite predominantly composed of nickel–iron.

Kepler's laws
The three laws of planetary motion, which describe the properties of an elliptical orbit produced by the gravitational interaction of two objects.

light curve
The variation of a meteor's brightness with time.

limiting magnitude
The magnitude of the faintest star visible.

line density
The number of electrons per metre length of meteor train column.

longitude of the ascending node (Ω)
The angle in the plane of the ecliptic between the vernal equinox and the line of nodes.

maximum
The time during a meteor shower at which the greatest numbers of meteors per hour are recorded.

meteor
The 'visible' phenomenon associated with the ablation of a meteoroid in the Earth's atmosphere.

meteorite
A meteoroid that has survived its passage through the Earth's atmosphere and landed on the Earth's surface.

Meteorite Observation and Recovery Project (MORP)
A project which ran a collection of twelve camera systems spread across the Prairie Provinces of Canada and gathered fireball data from 1971 to 1985.

meteoroid
Any object, less than 5m across, capable of producing a meteor upon entering the Earth's atmosphere.

meteoroid stream
A large collection of meteoroids, all ejected

from the same parent comet, moving about the Sun along similar orbits.

meteor storm
An event in which the *zenithal hourly rate* of a meteor shower is extremely high (by some definitions, over 1,000; by others, over 3,600).

nodes
The two points at which the orbit of a planet, comet or asteroid passes through the ecliptic: up through the ascending node and down through the descending node. The line joining the two nodes is called the line of nodes.

outburst
A rise in a meteor shower's zenithal hourly rate at maximum to at least two or three times the maximum rate recorded in typical years.

overdense
Describing a meteor train whose electron *line density* is sufficiently high for any incident radio wave to be totally reflected.

pallasite
A type of stony-iron meteorite consisting of nearly equal proportions of nickel–iron and olivine.

perihelion
The point at which an object, be it a planet, a comet, an asteroid or a meteoroid, in its orbit about the Sun is at its least distance from the Sun.

periodic shower
A meteor shower that undergoes strong outbursts of activity at intervals corresponding to the orbital period of its parent comet.

photon
A discrete 'package' of electromagnetic energy.

pixel
A picture element – the individual segments of a charge coupled device (CCD). The pixels in a CCD are arranged in a rectangular array and are read out sequentially in order to build up an electronic image.

plasma
See ionization.

radiant
The small region on the sky from which a particular meteor shower appears to emanate.

recombination
The reverse of ionization: the capture of an electron by an ion.

regmaglypts
Characteristic dimples produced on the surface of meteorites during their atmospheric flight.

right ascension (RA)
The east–west coordinate on the celestial sphere (similar to terrestrial longitude), measured anticlockwise around the celestial equator from the location of the spring equinox. It can be measured from 0 to 24 hours or (as in this book) from 0° to 360°.

semi-major axis
Half the longest diameter of an ellipse (or of an elliptical orbit).

solar longitude ($\lambda_{\odot}$)
The position of the Earth in its orbit measured as an angle along the ecliptic from the vernal equinox.

spectroscope
An instrument used to split light into its component colours (or individual wavelengths).

sporadic meteor
A meteor that cannot be linked to an active shower radiant.

strewnfield
The elliptical region over which fragments of the same meteorite fall are found.

sublimation
The transformation of a solid to a gas without an intervening liquid stage. Cometary ices, for example, sublimate from the solid water ice phase to a gas phase when they are heated by the Sun.

underdense
Describing a meteor train whose electron *line density* is small enough for any incident radio wave to be scattered by each individual electron in the train.

VLF
Very low frequency radiation, or equivalently very long wavelength radiation.

wavelength
The distance between two successive wave crests.

Widmanstätten pattern
The regular, hatched intergrowth of nickel–iron crystal structures in some iron meteorites.

zenith
The point on the celestial sphere directly overhead for an observer.

zenith angle
The angular distance from the start of a meteor trail to the observer's zenith.

zenithal hourly rate (ZHR)
The number of meteors a visual observer would see under perfect viewing conditions if the shower radiant were directly overhead.

Bibliography

Further Reading

Many excellent texts on meteor astronomy and the study of meteorites have been written over the years. Below are a few of the books that have inspired me, and while a number of them may now be considered somewhat out of date, they have been written by masters in the field and are still well worth reading.

Bone, Neil, Meteors: *A Sky and Telescope Guide* (London, George Philip, 1992; Cambridge, Mass., Sky Publishing, 1993)

Brandt, John C., *Introduction to Comets* (Cambridge University Press, 2004)

Brandt, John C. and Chapman, Robert D., *Rendezvous in Space: The Science of Comets* (New York: W. H. Freeman, 1992)

Burke, John G., *Cosmic Debris: Meteorites in History* (Berkeley and Los Angeles: The University of California Press, 1986)

Cassidy, William A., *Meteorites, Ice, and Antarctica: A Personal Account* (Cambridge University Press, 2003)

Clube, Victor and Napier, Bill, *The Cosmic Winter* (Oxford: Basil Blackwell, 1990)

Dodd, Robert T., *Thunderstones and Shooting Stars: The Meaning of Meteorites* (Cambridge, Mass: Harvard University Press, 1986)

Hawkins, Gerald S., *Meteors, Comets, and Meteorites* (New York: McGraw-Hill, 1964)

Kronk, Gary W., *Meteor Showers: A Descriptive Catalog* (Hillside, N.J.: Enslow Press, 1988) [An updated version of this text can be accessed on line at http://comets.amsmeteors.org/meteors/calendar.html]

Littmann, Mark, *The Heavens on Fire: The Great Leonid Meteor Storms* (Cambridge University Press, 1998)

Lovell, A.C.B., *Meteor Astronomy* (London: Oxford University Press, 1954)

McKinley, Donald W.R., *Meteor Science and Engineering* (New York: McGraw-Hill Publishing, 1961)

McSween, Harry Y. Jr, *Meteorites and Their Parent Planets*, second edition (Cambridge University Press, 1999)

Murad, Edmond and Williams, Iwan P. (editors), *Meteors in the Earth's Atmosphere* (Cambridge University Press, 2002)

Norton, O. Richard, *Rocks from Space: Meteorites and Meteorite Hunters*, second edition (Missoula, Mont.: Mountain Press Publishing, 1998)

Norton, O. Richard, *The Cambridge*

Encyclopedia of Meteorites (Cambridge University Press, 2002)

Olivier, Charles P., *Meteors* (London: Williams & Wilkins, 1925)

Olson, Roberta J. M. and Pasachoff, Jay M. *Fire in the Sky: The Decisive Centuries, in British Art and Science* (Cambridge University Press, 1998)

Opik, Ernst J., *The Physics of Meteor Flight in the Atmosphere* (New York: Interscience, 1958; reprint New York: Dover, 2004)

Porter, John G., *Comets and Meteor Streams* (New York: Wiley, 1952)

Rendtel, Jurgen, Arlt, Rainer and McBeath, Alastair (editors), *Handbook for Visual Meteor Observers*, IMO Monograph No. 2 (Potsdam, International Meteor Organization, 1995)

Steel, Duncan, *Target Earth* (Pleasantville, N.Y.: Readers Digest Association, 2000; London: Time Life Books, 2001)

Tirion, Wil and Sinnott, Roger, *Sky Atlas 2000.0*, laminated deluxe edition (Cambridge, Mass., Sky Publishing, 1999)

Yeomans, Donald K., *Comets: A Chronological History of Observations, Science, Myth, and Folklore* (New York: John Wiley, 1991)

Zanada, Brigitte and Rotaru, Monica, *Meteorites: Their Impact on Science and History* (Cambridge University Press, 2001)

Index